四川省示范性高职院校建设项目成果
校企合作共同编写，与企业对接，实用性强

网络互联实践

Wangluo Hulian Shijian

主　编◎周瑾怡　王宏旭
副主编◎张倩莉　陈新华
主　审◎王宇峰

西南交通大学出版社

图书在版编目（CIP）数据

网络互联实践 / 周瑾怡，王宏旭主编. —成都：
西南交通大学出版社，2015.8
ISBN 978-7-5643-4242-5

Ⅰ. ①网… Ⅱ. ①周… ②王… Ⅲ. ①互联网络 – 高等职业教育 – 教材 Ⅳ. ①TP393.4

中国版本图书馆 CIP 数据核字（2015）第 204046 号

网络互联实践
主编　周瑾怡　王宏旭

责 任 编 辑	李芳芳
特 邀 编 辑	韩迎春　林 莉
封 面 设 计	迷迦设计工作室
出 版 发 行	西南交通大学出版社 （四川省成都市金牛区交大路 146 号）
发 行 部 电 话	028-87600564　028-87600533
邮 政 编 码	610031
网　　　　址	http://www.xnjdcbs.com
印　　　　刷	成都中铁二局永经堂印务有限责任公司
成 品 尺 寸	185 mm × 260 mm
印　　　　张	8.75
字　　　　数	217 千
版　　　　次	2015 年 8 月第 1 版
印　　　　次	2015 年 8 月第 1 次
书　　　　号	ISBN 978-7-5643-4242-5
定　　　　价	20.00 元

课件咨询电话：028-87600533
图书如有印装质量问题　本社负责退换
版权所有　盗版必究　举报电话：028-87600562

序

2014年6月23至24日,全国第七次职业教育工作会议在北京召开,中共中央总书记、国家主席、中央军委主席习近平就加快职业教育发展作出重要指示。他强调,职业教育是国民教育体系和人力资源开发的重要组成部分,是广大青年打开通往成功成才大门的重要途径,肩负着培养多样化人才、传承技术技能、促进就业创业的重要职责,必须高度重视、加快发展。

在国家大力发展职业教育、创新人才培养模式的新形势下,加强高职院校教材建设及课程资源建设,是深化教育教学改革和全面培养技术技能人才的前提和基础。

近年来,四川信息职业技术学院坚持走"根植信息产业、服务信息社会"的特色发展之路,始终致力于打造西部电子信息高端技术技能人才培养高地,立志为电子信息产业和区域经济社会发展培养技术技能人才。在省级示范性高等职业院校建设过程中,学院通过联合企业全程参与教材开发与课程建设,组织编写了涉及应用电子技术、软件技术、计算机网络技术、数控技术四个示范建设专业的具有较强指导作用和较高现实价值的系列教材。

在编著过程中,编著者基于"理实一体"、"教学做一体化"的基本要求,秉承新颖性、实用性、开放性的基本原则,以校企联合为依托,基于工作过程系统化课程开发理念,精心选取教学内容、优化设计学习情境,最终形成了这套示范系列教材。本套教材充分体现了"企业全程参与教材开发、课程内容与职业标准对接、教学过程与生产过程对接"的基本特点,具体表现在:

一是编写队伍体现了"校企联合、专兼结合"。教材以适应技术技能人才培养为需求,联合四川军工集团零八一电子集团、联想集团、四川长征机床集团有限公司、宝鸡机床集团有限公司等知名企业全程参与教材开发,编写队伍既有企业一线技术工程师,又有学校的教授、副教授,专兼搭配。他们既熟悉国家职业教育形势和政策,又了解社会和行业需求;既懂得教育教学规律,又深谙学生心理。

二是内容选取体现了"对接标准,立足岗位"。教材编写以国家职业标准、行业标准为指南,有机融入了电子信息产业链上的生产制造类企业、系统集成企业、应用维护企业或单位的相关技术岗位的知识技能要求,使课程内容与国家职业标准和行业企业标准有机融合,学生通过学习和实践,能实现从学习者向从业者能力的递进。突出了课程内容与职业标准对接,使教材既可以作为学校教学使用,也可作为企业员工培训使用。

三是内容组织体现了"项目导向、任务驱动"。教材基于工作过程系统化理念开发，采用"项目导向、任务驱动"方式组织内容，以完成实际工作中的真实项目或教学迁移项目为目标，通过核心任务驱动教学。教学内容融基础理论、实验、实训于一体，注重培养学生安全意识、团队意识、创新意识和成本意识，做到了素质并重，能让学生在模拟真实的工作环境中学习和实践，突出了教学过程与生产过程对接。

四是配套资源体现了"丰富多样、自主学习"。本套教材建设有配套的精品资源共享课程（见 http://www.scitc.com.cn/），配置教学文档库、课件库、素材库、习题及试题库、技术资料库、工程案例库，形成了立体化、资源化、网络化的开放式学习平台。

尽管本套教材在探索创新中还存在有待进一步提升之处，但仍不失为一套针对高职电子信息类专业的好教材，值得推广使用。

此为序。

四川省高职高专院校
人才培养工作委员会主任 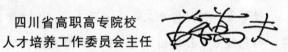

前　言

随着互联网的迅猛发展，为了满足社会发展与人们工作和生活的需要，网络信息化是一种必然发展趋势，社会对计算机网络技术专业人才的需求也与日俱增。然而很多计算机网络技术专业的学生只具备一些网络的理论知识，对实际网络的设备配置却非常生疏甚至完全不懂，为此我们编写了以操作实践为主、结合实际网络应用的实践教程。全书共分五大部分：

交换实践篇：主要介绍交换机的基本配置、VLAN 划分、生成树协议配置、VRRP、交换机端口聚合等实践项目。

路由实践项目：主要介绍路由器基本配置、静态路由配置、动态路由配置、路由重分布等实践项目。

网络安全实践篇：主要介绍在企业网内基于 IP 的 ACL 配置实践项目。

IP 地址服务实践篇：主要介绍静态 NAT 和 PAT 配置应用。

综合实践篇：通过实训室网络设备搭建企业网络环境，完成综合网络应用案例。

本书以思科网络设备命令行为蓝本进行编写，每个实训内容都是以项目形式呈现的，各项目包括项目名称、工程目标、技术要点、设备清单、工作场景、地址规划、工作过程等，而且实验内容与实际网络应用紧密结合，让学生能清晰、完整地完成各个实验项目。

本书由四川信息职业技术学院周瑾怡、王宏旭、张倩莉、陈新华及成都福立盟有限公司的王宇峰共同完成编写，其中王宏旭编写项目 1~5，陈新华编写项目 6~10、17，张倩莉编写项目 11~13，周瑾怡完成项目 14~16、18~22 的编写及全书的统稿工作，王宇峰对全书进行了修改和审核。

本书由校企合作共同完成，在整个编写过程中得到了四川信息职业技术学院、成都福立盟有限公司及思科网络学院的大力支持和帮助。本书在筹划、编写阶段还得到了赵克林、朱龙等几位教授的指导，在此一并表示由衷的感谢。

由于编写时间仓促，加之作者水平有限，书中难免有不妥和疏漏之处，恳请同行专家及读者指正。有任何意见或建议请联系 E-mail:30940350@qq.com。

编　者

2015 年 5 月

目　　录

第 1 章　交换实践篇 1
- 项目 1　交换机基本配置 1
- 项目 2　交换机远程管理配置 8
- 项目 3　交换机端口安全管理配置 11
- 项目 4　虚拟局域网 VLAN 配置 16
- 项目 5　VLAN 中继配置 22
- 项目 6　链路聚合配置 28
- 项目 7　生成树协议（STP）配置 32
- 项目 8　快速生成树（RSTP）配置 42
- 项目 9　VLAN 间路由配置 49

第 2 章　路由实践篇 53
- 项目 10　路由器基本配置 53
- 项目 11　单臂路由实现 VLAN 间路由 58
- 项目 12　静态路由配置 62
- 项目 13　默认路由配置 68
- 项目 14　RIP 路由协议应用 73
- 项目 15　配置 OSPF 路由实现区域网连通 79
- 项目 16　路由重分布 84
- 项目 17　虚拟路由器冗余协议（VRRP）配置 93

第 3 章　网络安全实践篇 99
- 项目 18　配置标准访问列表控制网络流量 99
- 项目 19　配置扩展访问列表保护服务器安全 103

第 4 章　IP 地址服务实践篇 107
- 项目 20　静态 NAT 配置 107
- 项目 21　端口多路复用（PAT）配置 112
- 项目 22　动态主机配置协议（DHCP）配置 116
- 项目 23　DHCP 中继配置 120

第 5 章　综合实践篇 124
- 项目 24　中小企业园区网建设 124

第1章 交换实践篇

项目1 交换机基本配置

1. 工程目标

（1）了解交换机的工作原理。
（2）掌握交换机基本配置方法。

2. 技术要点

1）交换机的组成

交换机的结构与计算机相似，都由 CPU、存储器、输入/输出设备、操作系统等组成，其交换功能也是由硬件和软件共同实现的。交换机硬件主要包括 CPU、内存、主板、端口等，其端口类型主要有以太网端口、快速以太网端口、吉比特以太网端口和控制台端口。其存储介质主要有只读存储器（ROM）、闪存（FLASH）、非易失性存储器（Non-Volatile RAM）和动态随机存储器（Dynamic RAM）。交换机软件主要包括交换机操作系统、交换机配置文件等。

2）交换机的启动

交换机通电后，首先运行 ROM 中的自检程序，进行系统自检。自检通过后，系统会引导运行 FLASH 中的 IOS 操作系统，进行操作系统加载。加载系统后会运行 NVRAM 中的交换机配置文件，将其装入 DRAM 中，进行配置文件加载，成功后则完成交换机启动。

3）交换机的工作原理

交换机作为数据链路层的设备，其主要功能是对计算机网络数据帧进行相应操作和管理。交换机根据数据包中的 MAC 地址进行学习、转发和过滤，从而实现数据的传输。

MAC 地址学习：交换机在转发数据时，以 MAC 地址表为依据，其 MAC 地址表主要由与本交换机相连的终端主机的 MAC 地址以及本交换机连接主机的端口等信息组成。当交换机刚刚启动时，其 MAC 地址表中的内容为空，这时如果交换机的某个端口收到数据帧，它会把数据帧泛洪到除接收端口外的所有端口，使网络中其他所有的终端主机都能收到此数据帧，同时交换机通过端口记录下相关数据帧的 MAC 地址，将其填充到 MAC 地址表中，完成 MAC 地址和端口的对应关系，实现交换机的 MAC 地址学习。另外由于交换机存储空间有限，

MAC 地址不能无休止地学习，为防止 MAC 地址数量超出存储容量，交换机在 MAC 地址表中为每条表项设定一个老化时间，如果在老化时间到期之前一直没有刷新，则 MAC 表将会清除该表项，为新的 MAC 地址腾出存储空间。

MAC 地址转发：在 MAC 地址表初始化后，交换机就会根据 MAC 地址表项中的内容进行数据帧转发。其转发规则为：对于已知数据帧（目的 MAC 地址在交换机 MAC 地址表中有相应表项），则根据表中内容将数据帧通过相对应的端口转发出去；对于未知单播数据帧（目的 MAC 地址在交换机 MAC 地址表中没有相应表项）、组播帧和广播帧，则采用泛洪的方式将数据帧从除接收端口外的所有端口转发出去。

MAC 地址过滤：为了杜绝不必要的帧转发，交换机对符合特定条件的帧进行过滤。如果帧目的 MAC 地址在 MAC 地址表中有表项存在，且表项所关联的端口与接收到帧的端口相同，则交换机对此数据帧进行过滤，即不转发此数据帧。

4）交换机的分类

交换机作为数据网络传输设备，其转发数据帧的模式有三种：直通转发、存储转发和无碎片转发，其中存储转发是交换机的主流交换方式。

- 直通转发。

直通转发是指交换机在接收端口检测到一个数据帧时，立刻检查该数据帧的帧头，获取目的 MAC 地址，根据交换机 MAC 表中的表项内容将该数据帧转发。

- 存储转发。

存储转发是指交换机会从接收端口中完整地接收整个数据帧，读取目的 MAC 地址和源 MAC 地址，进行运算比对后才进行数据帧的转发。

- 无碎片转发。

无碎片式交换方式介于直通转发和存储转发之间，转发原则是交换机读取前 64 个字节后开始转发。该交换方式的数据处理速度比存储转发方式快，比直通转发方式慢，但由于它能够避免残帧的转发，所以被广泛地应用于低档交换机。

3. 设备清单

交换机（1 台）；配置线缆（1 条）；网络连线（若干根）；测试 PC（1 台）；备份服务器（1 台）。

4. 工作场景

小王是信息学院网络信息中心新进的网管，负责网络中心的设备管理工作。来学校网络信息中心报到后，学校网络信息中心要求小王熟悉学校安装的网络产品。学校网络中心采用的是全系列思科网络产品，现要求小王登录配置交换机，了解、掌握交换机的命令行操作。

交换机基本配置工作场景如图 1.1 所示，主要设备有：1 台思科二层交换机、1 台 PC 终端和 1 台 TFTP 服务器。PC 终端分别通过配置线和网线（F0/1 端口）连接到交换机上，TFTP 服务器通过网线连接到交换机（F0/24 端口）上，通过 PC 终端对交换机进行基础配置和管理。

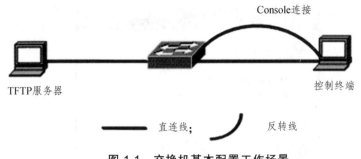

图 1.1 交换机基本配置工作场景

5. 地址规划

设备网络地址规划表如表 1.1 所示。

表 1.1 设备网络地址规划表

设备名称	设备及端口的配置地址		备注
交换机	F0/1	192.168.1.1/24	局域网端口，连接控制终端 PC1
	F0/24	192.168.1.24/24	局域网端口，连接 TFTP 服务器
控制终端 PC1	192.168.1.2/24		
TFTP 服务器	192.168.1.254/24		
特权密码	123		
控制台密码	123		

6. 工作过程

步骤一：连接设备。如图 1.1 所示使用网络线在工作现场连接好设备。

交换机、路由器等网络设备的管理方式可以简单地分为两种：带外管理和带内管理。所谓带内管理，是指网络的管理控制信息与用户网络的承载业务信息通过同一个逻辑信道传送，占用业务带宽；而带外管理是指网络的管理控制信息与用户网络的承载业务信息在不同的逻辑信道传送，不占用业务带宽。

通过 Console 口管理是最常用的带外管理方式。通常用户会在首次配置交换机或者无法进行带内管理时使用带外管理方式。交换机等网络设备的配置线一般都是设备自带，通常一端为 RJ-45 接口，另一端为 DB-9 端口。

正确将配置线连接到 PC 和交换机后，通过 PC 自带的超级终端实现对交换机的初始化配置。具体配置步骤如下：

（1）单击"开始"→"程序"→"附件"→"通信"→"超级终端"。

（2）打开超级终端，输入名称；选择用以连接交换机的 COM 口编号；设置端口属性（波特率为 9 600 bps，数据位为 8，奇偶校验无，停止位为 1，数据流控制为无）如图 1.2～1.4 所示。

（3）进入交换机配置界面，进行交换机初始化配置。

图 1.2　超级终端配置（1）

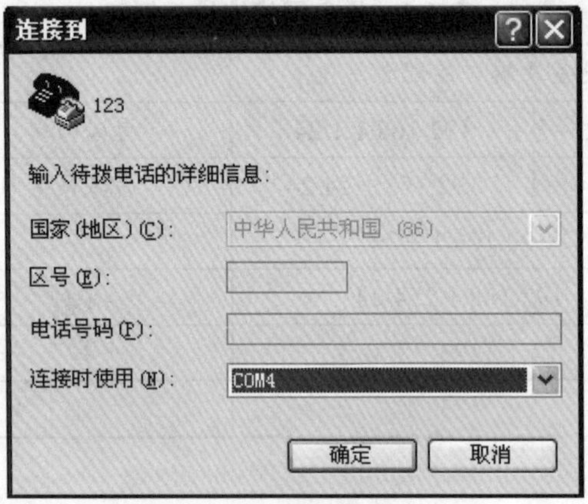

图 1.3　超级终端配置（2）

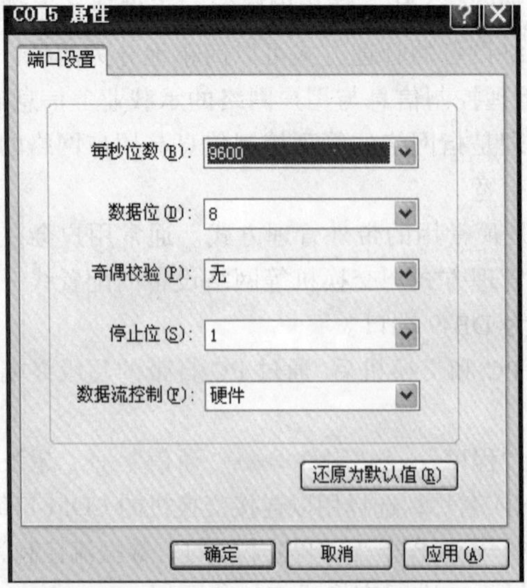

图 1.4　超级终端配置（3）

步骤二：配置交换机设备基本信息。

（1）交换机基本配置模式。

为了方便管理交换机以及相关安全方面的考虑，将配置模式进行了细分，一般交换机提供的有用户 EXEC 模式、特权 EXEC 模式、全局配置模式、接口配置模式、VLAN 数据库配置模式等多种级别。

① 用户 EXEC 模式。

用户通过交换机的控制台端口或 Telnet 端口连接登录到交换机时，进入的模式即为用户 EXEC 模式。该模式下，用户大多只能使用一些基本的查看、设置和测试命令。

用户 EXEC 模式命令状态行为：switch>。

② 特权 EXEC 模式。

在用户模式下输入 enable 命令，即可进入到特权 EXEC 模式。在该模式下用户获得了更高的系统使用权限，可以使用更多的系统命令对系统进行操作。

特权 EXEC 模式命令状态行为：switch#。

③ 全局配置模式。

在特权模式下输入 configure terminal 命令，即可进入全局配置模式。该模式的命令主要是对交换机全局进行管理，对整个交换机起作用。

全局配置模式命令状态行为：switch(config)#。

④ 接口配置模式。

在全局模式下输入 interface 命令，即可进入接口配置模式，该模式的命令主要是对选定的交换机接口进行配置，更改接口属性。

接口配置模式命令状态行为：switch(config-if)#。

主要命令模式介绍如表 1.2 所示。

表 1.2 主要命令模式介绍表

命令模式	访问方法	提示符	离开或访问其他模式方法
用户 EXEC 模式	访问交换机时首先进入该模式	switch>	输入 exit 命令离开该模式。输入 enable 命令进入特权模式
特权 EXEC 模式	在用户模式下，使用 enable 命令进入该模式	switch#	输入 disable 命令返回用户模式，输入 configure terminal 命令进入全局配置模式
全局配置模式	在特权模式下，使用 configure terminal 命令进入该模式	switch(config)#	输入 exit 命令返回到特权模式
接口配置模式	在全局配置模式下，使用 interface 命令进入该模式	switch(config-if)#	输入 end 命令返回到特权模式，输入 exit 返回到全局配置模式，在 interface 命令中必须指明要进入哪一个接口配置子模式

（2）配置交换机名字。

在日常生活中，为了方便识别交换机，网络管理员往往给所管理的交换机起名字，例如：

```
Switch>
Switch>enable
Switch#configure terminal
Enter configuration commands, one per line.   End with CNTL/Z.
Switch(config)#hostname sw1
sw1(config)#
```

（3）配置交换机的终端密码。

为了保障交换机的使用安全，网络管理员往往要给所管理的交换机设置密码。

① 给交换机 console 设置密码，例如：

```
sw1>
sw1>enable                              //进入特权模式
sw1#configure terminal                  //进入全局配置模式
sw1(config)#line console 0              //进入配置端口 console 口
sw1(config-line)#password 123           //配置端口 console 口密码为 123
sw1(config-line)#login
sw1(config-line)#exit
sw1(config)#
```

② 给交换机虚拟终端设置密码，例如：

```
sw1>
sw1>enable                              //进入特权模式
sw1#configure terminal                  //进入全局配置模式
sw1(config)#line vty 0 15               //配置 vty 接入 0～15
sw1(config-line)#password 123           //配置 vty 连接密码为 123
sw1(config-line)#login                  //使用 login 命令配置使用
sw1(config-line)#exit
```

③ 给交换机特权模式设置密码，例如：

```
sw1#
sw1#configure terminal                  //进入全局配置模式
sw1(config)#enable password 123         //配置 enable 连接密码为 123
```

（4）保存交换机配置文件。

为保障网络通信能正常进行，网络管理员应对交换机配置进行备份，而通常的做法是网络管理员在正确配置交换机后，将交换机的配置文件从交换机上下载并保存在稳妥的地方，当交换机出现故障后可以将备份的配置文件重新导入，从而减少了重新配置的麻烦，缩短了故障排除时间。同样，网络管理员有时也将交换机的 IOS 文件进行备份，以备交换机 IOS 故障时可以进行恢复。

目前，较为常用的一种备份和恢复交换机数据的方法是采用 TFTP 服务器，在网络中设置一个TFTP服务器,然后通过网络将服务器与交换机相连,在交换机上使用copy命令将flash文件、config文件拷贝到服务器上或从服务器上将文件拷贝到交换机上,从而实现交换机上的文件备份和还原。

TFTP服务器设置十分简单,只需在网上下载一个TFTP服务器安装软件,点击安装后进行日志文件和根目录设置即可使用,特点是功能精简、小巧灵活,如图1.5所示。

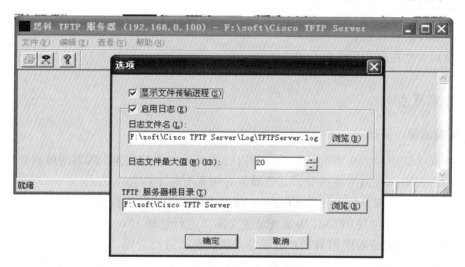

图 1.5 TFTP 服务器配置图

交换机配置文件备份是将交换机的当前运行配置文件 running-config 或启动配置文件 startup-config 保存到 TFTP 服务器,从而实现备份。交换机配置文件恢复是从 TFTP 服务器上下载以前备份的文件到交换机上,作为启动配置文件。

① 保存当前配置文件:

sw1#copy running-config startup-config

② 保存当前配置文件到 TFTP 服务器:

sw1#copy running-config tftp

Address or name of remote host []? 192.168.1.254

Destination filename [sw1-config]? sw1

7. 总　　结

交换机作为网络传输设备主要工作在数据链路层,它通过 MAC 地址表实现数据帧的传输。交换机的内部结构与计算机的十分相似,其功能的实现主要基于 IOS 操作系统。交换机在使用前要通过 Console 口对其进行初始化配置,初始化后即可在交换机各种模式下对其进行配置管理。

项目 2　交换机远程管理配置

1. 工程目标

（1）了解交换机远程管理方法。
（2）掌握交换机远程管理配置命令。

2. 技术要点

1）交换机带内管理

在日常网络管理中，交换机通常采用带内管理的方法进行配置。所谓的带内管理是指交换机管理时占用交换机的网络带宽。其管理方式主要有以下四种：
（1）通过 Telnet 客户软件使用 TELNET 协议登录到交换机进行管理。
（2）通过 SSH 软件使用 SSH 协议登录到交换机进行管理。
（3）通过 Web 浏览器使用 HTTP 协议登录到交换机进行管理。
（4）通过网络管理软件（如 Cisco Works）使用 SNMP 协议对交换机进行管理。

带内管理是指计算机通过网络的方式连接到交换机上，从而便于网络管理人员从远程登录到交换机上进行管理，因此要在网络中找到交换机就必须要给交换机配置一个用于网络管理的 IP 地址，否则管理设备无法在网络中定位寻找到被管理的交换机。当交换机的配置出现变更，导致带内管理失效时，必须使用带外管理对交换机进行配置管理。

2）交换机安全管理

交换机是网络中的数据传输设备，它和许多 PC 机、服务器、路由器等网络设备相连，转发大量的数据，可以说它是网络传输的中枢。因此，交换机的管理十分重要。网络管理员为保证交换机配置内容不会轻易地被攻击修改，就需对交换机的管理进行安全配置，以保证其安全运行。交换机配置管理安全图如图 1.6 所示。

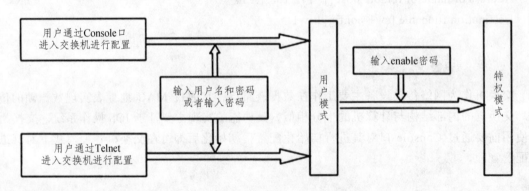

图 1.6　交换机配置管理安全图

3. 设备清单

交换机（1台）；网络连线（若干根）；测试PC（1台）。

4. 工作场景

小王是广东某市外语学校网络中心新进的网管，负责网络中心的设备管理工作。来学校网络中心报到后，学校网络中心要求小王熟悉学校安装的网络产品。交换机基本配置工作场景如图1.7所示，主要设备有：1台思科二层交换机、1台PC终端，PC终端分别通过配置线和网线（f0/1端口）连接到交换机上，PC终端通过网线对交换机进行基础配置和管理。

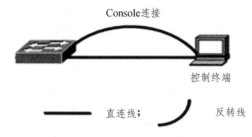

图1.7 交换机基本配置工作场景

5. 地址规划

设备网络地址规划如表1.3所示。

表1.3 设备网络地址规划表

设备名称	设备配置地址	备 注
交换机	192.168.1.1/24	VLAN1
PC1	192.168.1.2/24	局域网端口 f0/1 连接 PC1

6. 工作过程

步骤一：利用配置线缆与交换机上Console口相连。
（1）配置交换机Telnet登录。

Switch>enable
Switch#configure terminal
switch(config)#enable password 123   //设置特权模式口令
switch(config)#line vty 0 15   //设置虚拟终端
switch(config-line)#password 456   //设置虚拟终端登录用户口令为456
switch(config-line)#login   //在线路配置模式下使用login命令测试
switch(config-line)#

注：如果没有出现提示，说明设置成功，如果出现五行提示，说明设置有误。

（2）配置交换机 IP 地址。

switch(config)#interface   vlan 1   //进入 vlan 1 接口模式
switch(config-if)#ip address 192.168.1.1 255.255.255.0   //配置 vlan 1 的 IP 地址
switch(config-if)#no shutdown   //打开 vlan 1 端口

（3）使用 PC1 机远程登录交换机。
① 将 PC1 用直通线连接到交换机的 FastEthernet0/1 端口。
② 设置 PC1 静态配置 IP 地址为 192.168.1.2，子网掩码为 255.255.255.0，如图 1.8 所示。

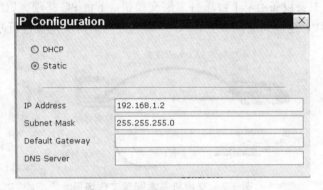

图 1.8　PC 端 IP 地址配置图

③ 在 PC1 上运行"命令提示符"，使用 Telnet 命令远程登录交换机，如图 1.9 所示。

PC>telnet 192.168.1.1
　　Password:password（输入密码为前面交换机上设定的 Telnet 登录口令 456）

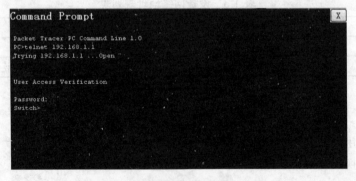

图 1.9　PC 端远程登录交换机配置图

7. 总　　结

本节主要讲授的是交换机的远程管理。日常生活中交换机的管理方式大多采用远程网络管理，管理员通过 Telnet 实现交换机的网络连接，采用远程登录的方式对交换机进行配置，操作起来方便快捷。在实现远程管理前一定要对交换机进行密码设置和 IP 地址设置，以提高交换机的安全性，实现交换机的网络可查找。

项目 3 交换机端口安全管理配置

1. 工程目标

（1）了解交换机端口管理方法。
（2）掌握交换机端口管理配置命令。

2. 技术要点

交换机端口安全功能是指针对交换机的端口进行安全属性的配置，从而控制用户的安全接入。交换机端口安全配置主要包括两项内容：一是限制交换机端口的最大连接数，二是针对交换机端口进行 MAC 地址、IP 地址的绑定。

限制交换机端口的最大连接数可以控制交换机端口下连的主机数，并防止用户进行恶意的 ARP 欺骗。

交换机端口的地址绑定，可以针对 IP 地址、MAC 地址、IP + MAC 进行灵活的绑定，实现对用户的严格控制，保证用户的安全接入和防止常见的内网的网络攻击，如 ARP 欺骗，IP、MAC 地址欺骗，IP 地址攻击等。

配置了交换机的端口安全功能后，当实际应用超出配置的要求时，将产生一个安全违例，产生安全违例的处理方式有三种：

Protect：当安全地址个数满后，安全端口将丢弃未知名地址（不是该端口的安全地址中的任何一个）的包。

Restrict：当违例产生时，将发送一个 Trap 通知。

Shutdown：当违例产生时，将关闭端口并发送一个 Trap 通知。

当端口因为违例而被关闭后，在全局配置模式下使用命令 errdisable recovery 来将接口从错误状态中恢复过来。

3. 设备清单

交换机（1 台）；网络连线（若干根）；测试 PC（1 台）。

4. 工作场景

某公司在内部的交换机上配置每个端口可连接的最大安全地址数，可以防止外来用户盗用公司内部网络。公司的主交换机上绑定公司内部主机的 MAC 地址之后，能防止公司外部带有病毒的 PC 机感染公司内部主机，或者防止人为的恶意破坏，窃取公司内部机密。

交换机基本配置工作场景如图 1.10 所示，主要设备有：1 台思科二层交换机、1 台 PC

终端,PC 终端分别通过配置线和网线(f0/1 端口)连接到交换机上,PC 终端通过网线对交换机进行基础配置和管理。

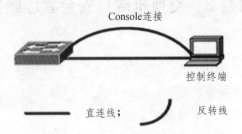

图 1.10　交换机基本配置工作场景

5. 地址规划

设备网络地址规划表如表 1.4 所示。

表 1.4　设备网络地址规划表

设备名称	设备配置地址	备　注
交换机	192.168.1.1/24	VLAN1
PC1	192.168.1.2/24	MAC:0090.0c61.3e70,与端口 f0/1

6. 工作过程

1)交换机的 MAC 地址表管理

① 查看主机的 MAC 地址。

在命令行下输入 ipconfig/all 命令查看主机的 MAC 地址并记录,如图 1.11 所示。

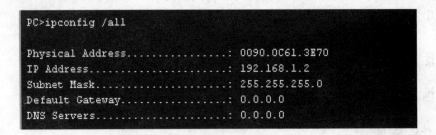

图 1.11　PC 机 MAC 地址查询图

② 查看交换机的 MAC 地址表(见图 1.12、1.13)。
Switch#show mac-address-table

```
Switch#show mac-address-table
          Mac Address Table
-------------------------------------------

Vlan     Mac Address       Type        Ports
----     -----------       --------    -----

 1       0090.0c61.3e70    DYNAMIC     Fa0/1
```

图 1.12 查看交换机上的 MAC 地址表

Switch#show mac-address-table static

```
Switch#show mac-address-table static
          Mac Address Table
-------------------------------------------

Vlan     Mac Address       Type        Ports
----     -----------       --------    -----
```

图 1.13 查看交换机上的静态 MAC 地址表

③ 清空交换机的 MAC 地址表（见图 1.14）。
Switch#clear mac-address-table dynamic
Switch#show mac-address-table

```
Switch#clear mac-address-table dynamic
Switch#show mac-address-table
          Mac Address Table
-------------------------------------------

Vlan     Mac Address       Type        Ports
----     -----------       --------    -----
```

图 1.14 交换机上清空 MAC 地址表

④ 添加静态 MAC 地址。

若要使某个 MAC 地址永久地与交换机的某一个端口相关联，可以手动添加 MAC 地址到交换机端口上，完成 MAC 地址和端口的映射，这个 MAC 地址被称为静态 MAC 地址。使用静态 MAC 地址的优点如下：

- 该 MAC 地址不会因为超时而被交换机自动清除。
- 网络上的服务器被要求连接到交换机指定端口。
- 便于网络管理，提高交换机与终端设备的通信安全性。

添加静态 MAC 地址如图 1.15 所示。

Switch(config)#mac-address-table static xxxx.xxxx xxxx vlan xinterface fastethernet slot/port

注：xxxx.xxxx xxxx 为主机的 MAC 地址，slot/port 为与主机相连的交换机的端口号。

```
Switch(config)#mac-address-table static 0090.0c61.3e70 vlan 1 interface f0/1
Switch(config)#exit
Switch#

            Switch#show mac-address-table st
                  Mac Address Table
            -------------------------------------------

            Vlan      Mac Address       Type        Ports
            ----      -----------       --------    -----

              1       0090.0c61.3e70    STATIC      Fa0/1
```

图 1.15　交换机上添加静态 MAC 地址并查看

⑤ 删除静态 MAC 地址。

Switch(config)# no mac-address-table static xxxx. xxxx xxxx vlan xinterface fastethernet slot/port

2）交换机端口

一个交换机(包括二层和三层)的接口类型可分为两大类：二层接口（二层交换机）和三层接口（三层交换机）。而二层接口又分为 Access 口模式和 Trunk 口模式。

Access 交换口：只有二层交换功能，用于管理物理接口和与之相关的第二层协议，但不处理路由和桥接。交换机的接口默认都是 Access 交换口。每个交换机的 Access 交换口只能属于一个 VLAN，且只能传输属于这个 VLAN 的数据帧。

Trunk 交换口：可传输属于多个 VLAN 的数据帧。默认情况下 Trunk 交换口传输所有 VLAN 的数据帧，交换机也可通过设置 VLAN 许可列表来限制 Trunk 口只传输哪些 VLAN 的数据帧。

① 配置端口最大安全地址数量。

Switch(config)#interface f0/1
Switch(config-if)#switchport mode access   //将端口模式设置成 access 模式
Switch(config-if)#switchport port-security   //打开端口安全功能
Switch(config-if)#switchport port-security maximum 1   //设置端口安全地址最大数
Switch(config-if)#sw port-security violation protect / shutdown   //设置处理违例方式

注：当端口因为违例进入 "err-disabled" 状态后，在全局模式下输入命令才能将端口恢复为 up 状态。

Switch(config)#errdisable recovery cause security-violation

② 配置端口最大安全地址。

Switch(config)#interface f0/1
Switch(config-if)#switchport mode access   //将端口模式设置成 access 模式
Switch(config-if)#switchport port-security   //打开端口安全功能
Switch(config-if)#switchport port-security maximum 1   //设置端口安全地址最大数

Switch(config-if)# switchport port-security violation shutdown   //设置处理违例方式
Switch(config-if)# switchport port-security mac-address xx.xx.xx   //设置 MAC 地址

③ 查看端口安全信息。

端口安全配置完成后，可以使用 show 命令进行查看。

Switch#show port-security interface f0/1   //查看端口状态
Switch#show port-security address   //查看端口安全地址
Switch#show port-security   //查看所有端口安全信息

7. 总　　结

本节主要介绍了交换机的端口安全管理。通过对 MAC 地址的管理，交换机可以维护并修改自己的 MAC 表项。通过对端口的安全设置，交换机可以限制连接端口的 PC 机数量、方式等，从而大大提高了交换机的接入安全。

项目 4 虚拟局域网 VLAN 配置

1. 工程目标

（1）了解 VLAN 的基本概念和作用。
（2）理解动态和静态 VLAN 的特点与区别。
（3）掌握静态 VLAN 的配置。

2. 技术要点

1）VLAN 概述

在以太网中，一台交换机的每个端口都是一个冲突域，从而降低了通信时的数据冲突，提高了数据转发效率。但是，由于交换机的工作原理，它的所有端口都处于一个广播域内，当交换机发送广播帧时，与交换机相连的设备主机都可以收到数据帧，而随着企业的发展及信息技术的普及，连接到交换机上的设备将会越来越多，由于处在同一个广播域，大量的广播报文将会带来带宽浪费以及通信安全问题。

虚拟局域网（VLAN）技术是将连在交换机上的主机按使用需求划分到不同的虚拟网段，将广播域进行切割，划分出数量多、规模相对较小的广播域，从而减少广播对网络的不利影响。

网络管理人员通过手工方式将交换机的不同端口标记为属于不同的 VLAN，将接入某端口中的主机将划分到 VLAN 上。

2）VLAN 分类

① 静态 VLAN。
静态 VLAN 是指网络管理员通过手工配置将交换机端口分配给某一个 VLAN。其特点是配置简单、安全、易于实现和监视。
② 动态 VLAN。
动态 VLAN 是指管理员建立一个较复杂的 VLAN 数据库，当设备连接交换机时，交换机按数据库内容自动把这个网络设备所连接的端口分配到相应的 VLAN 中。

按照使用需求的不同，虚拟局域网有以下几种：
- 基于端口的 VLAN；
- 基于协议的 VLAN；
- 基于 MAC 地址的 VLAN；
- 基于 IP 子网的 VLAN；

- 基于 IP 组播的 VLAN；
- 基于策略的 VLAN。

配置 VLAN 时需要注意：
- 不同交换机平台、不同 IOS 版本支持的 VLAN 最大数量不同。
- VLAN1 为管理 VLAN，它不能创建、删除或重命名。

③ Native VLAN。

所谓 Native VLAN，也叫默认 VLAN，在这个接口上收发未标记的报文，都被认为是属于 Native VLAN。通常 VLAN 1 作为默认的 Native VLAN 是无法删除。

3）交换机配置链路

① 接入链路（Access Link）。

接入链路是指用于连接主机和交换机的链路，该链路只能与单个 VLAN 相关。

② 干道链路（Trunk Link）。

干道链路是计算机之间点对点的链路，可以同时承载多个 VLAN 的通信量。

当一个未标记的帧经过 Trunk 口时，会被打上 Native VLAN 的标记；一个已标记的帧经过 Trunk 口时，如果其标记的 VIAN 与 Trunk 口的 Native VLAN 相同，则会被剥去标记。

一个交换机的端口若定义为 Access Port，在未将此端口分给任何 VLAN 时，默认情况下属于 VLAN 1，此时其 Native VLAN 也就是 VLAN 1。若将此端口划分给某个 VLAN，则此 VLAN 即为本端口的 Native VLAN，它只能传输属于这个 VLAN 的帧，其他帧不能传输。

3. 设备清单

交换机（4 台）；网络连线（若干根）；测试 PC（6 台）。

4. 工作场景

如图 1.16 所示的网络拓扑是中北某大学教学行政楼中多个教研组部门之间的网络连接场景。以前不同教研组之间的计算机都连接在同一台交换机上，网络中由于广播等干扰，传输效率低下。新规划的网络提出，希望通过实施虚拟局域网技术，保证不同教研组部门网络之间的计算机互相不进行干扰，实现隔离技术，提高网络传输效率。

主要设备有：2 台思科二层交换机、6 台 PC 终端。PC 终端分别通过网线连接到交换机上，2 台二层交换机彼此之间通过网线相连。PC 终端通过网线对交换机进行基础配置和管理。

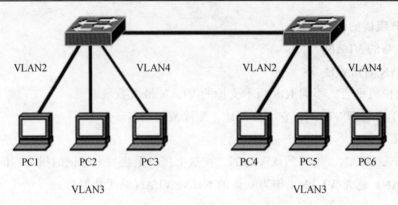

图 1.16 VLAN 配置工作场景

5. 地址规划

表 1.5 设备网络地址规划表

设备名称	IP/VLAN 设置	备注
交换机 1	VLAN2	划分 VLAN 端口 F0/2
	VLAN3	划分 VLAN 端口 F0/3
	VLAN4	划分 VLAN 端口 F0/4
交换机 2	VLAN2	划分 VLAN 端口 F0/2
	VLAN3	划分 VLAN 端口 F0/3
	VLAN4	划分 VLAN 端口 F0/4
PC1	192.168.2.1/24	连接端口 F0/2
PC2	192.168.3.1/24	连接端口 F0/3
PC3	192.168.4.1/24	连接端口 F0/4
PC4	192.168.2.2/24	连接端口 F0/2
PC5	192.168.3.2/24	连接端口 F0/3
PC6	192.168.4.1/24	连接端口 F0/4
交换机 1 与交换机 2 连接		端口 F0/24

6. 工作过程

步骤一：连接设备。如图 1.16 所示，使用网络线在工作现场连接好设备。

步骤二：配置交换机 VLAN。

（1）在交换机上创建 VLAN，如图 1.17 所示，再使用 show vlan 命令查看创建好的 VLAN，如图 1.18 所示。

```
Switch>enable
Switch#vlan database
Switch(vlan)#vlan 2 name vlan2          //将 VLAN 2 命名为 vlan2
Switch(vlan)#vlan 3 name vlan3          //将 VLAN 3 命名为 vlan3
Switch(vlan)#vlan 4 name vlan4          //将 VLAN 4 命名为 vlan4
```

```
Switch(vlan)#vlan 2 name vlan2
VLAN 2 added:
    Name: vlan2
Switch(vlan)#vlan 3 name vlan3
VLAN 3 added:
    Name: vlan3
Switch(vlan)#vlan 4 name vlan4
VLAN 4 added:
    Name: vlan4
```

图 1.17　交换机上创建 VLAN 基本配置工作场景

```
Switch#show vlan

VLAN Name                         Status    Ports
---- -------------------------- --------- -------------------------------
1    default                      active    Fa0/1, Fa0/2, Fa0/3, Fa0/4
                                            Fa0/5, Fa0/6, Fa0/7, Fa0/8
                                            Fa0/9, Fa0/10, Fa0/11, Fa0/12
                                            Fa0/13, Fa0/14, Fa0/15, Fa0/16
                                            Fa0/17, Fa0/18, Fa0/19, Fa0/20
                                            Fa0/21, Fa0/22, Fa0/23, Fa0/24
                                            Gig1/1, Gig1/2
2    vlan2                        active
3    vlan3                        active
4    vlan4                        active
1002 fddi-default                 act/unsup
1003 token-ring-default           act/unsup
1004 fddinet-default              act/unsup
1005 trnet-default                act/unsup
```

图 1.18　交换机上查看生成 VLAN

（2）将交换机 1 和交换机 2 上的端口划入 VLAN，使用 show vlan 命令查看结果，如图 1.19 所示。

```
Switch#configure terminal
Switch(config)#interface f0/2                   //进入 f0/2 接口
Switch(config-if)#switchport access vlan 2      //配置接口模式并将其划入 vlan 2 中
Switch(config-if)#no shutdown                   //启动 f0/2 接口
Switch(config)#interface f0/3                   //进入 f0/3 接口
Switch(config-if)#switchport access vlan 3      //配置接口模式并将其划入 vlan 3 中
Switch(config-if)#no shutdown                   //启动 f0/3 接口
Switch(config)#interface f0/4                   //进入 f0/4 接口
Switch(config-if)#switchport access vlan 4      //配置接口模式并将其划入 vlan 4 中
Switch(config-if)#no shutdown                   //启动 f0/4 接口
```

```
Switch#sho vlan

VLAN Name                             Status    Ports
---- -------------------------------- --------- -------------------------------
1    default                          active    Fa0/1, Fa0/5, Fa0/6, Fa0/7
                                                Fa0/8, Fa0/9, Fa0/10, Fa0/11
                                                Fa0/12, Fa0/13, Fa0/14, Fa0/15
                                                Fa0/16, Fa0/17, Fa0/18, Fa0/19
                                                Fa0/20, Fa0/21, Fa0/22, Fa0/23
                                                Fa0/24, Gig1/1, Gig1/2
2    vlan2                            active    Fa0/2
3    vlan3                            active    Fa0/3
4    vlan4                            active    Fa0/4
1002 fddi-default                     act/unsup
1003 token-ring-default               act/unsup
1004 fddinet-default                  act/unsup
1005 trnet-default                    act/unsup
```

图 1.19 交换机上查看 VLAN 端口划分

（3）跨交换机实现 VLAN 间的通信。

在同一台交换机上可以创建不同的 VLAN，每个 VLAN 是一个广播域，因而两个不同 VLAN 中主机相互不能通信。要实现跨交换机 VLAN 间的通信就需要在不同的交换机上创建相同的 VLAN，并在交换机之间建立 Trunk 链路，通过 Trunk 链路实现不同交换机相同 VLAN 的内部通信。

设置在交换机和交换机之间的连接端口 f0/24 模式：

Switch(config)#interface f0/24   //进入 f0/24 接口
Switch(config-if)#switchport mode trunk //配置接口模式为 trunk 模式

注意：每个 Trunk 口的缺省 Native VLAN 为 VLAN 1；在配置 Trunk 链路时，链路两端的 Trunk 口要属于相同的 Native VLAN。

（4）配置 PC1、PC2、PC3、PC4、PC5、PC6 的 IP 地址（以 PC1 为例）。

PC1 的 IP 配置如图 1.20 所示。

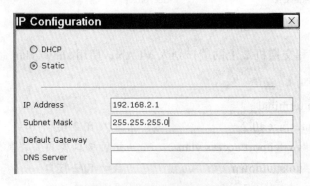

图 1.20 设置 PC 上的 IP 地址

（5）进行 PC 端的连通测试，如图 1.21 所示。

PC1—PC4 连通测试：在 PC1 端使用 ping 命令，测试与 PC4 是否连通。
PC2—PC5 连通测试：在 PC2 端使用 ping 命令，测试与 PC5 是否连通。
PC3—PC6 连通测试：在 PC3 端使用 ping 命令，测试与 PC6 是否连通。
命令：ping 192.168.2.2

注：因为 PC1 和 PC4 都在 VLAN 2 中，故 PC1 和 PC4 之间能进行通信；同理，PC2 和 PC5 都在 VLAN3 中，故 PC2 和 PC5 之间能进行通信；PC3 和 PC6 都在 VLAN 4 中，故 PC3 和 PC6 之间能进行通信。

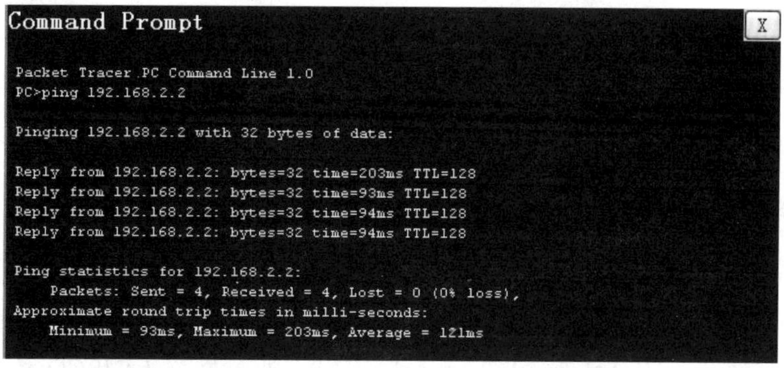

```
Command Prompt                                    X

Packet Tracer PC Command Line 1.0
PC>ping 192.168.2.2

Pinging 192.168.2.2 with 32 bytes of data:

Reply from 192.168.2.2: bytes=32 time=203ms TTL=128
Reply from 192.168.2.2: bytes=32 time=93ms TTL=128
Reply from 192.168.2.2: bytes=32 time=94ms TTL=128
Reply from 192.168.2.2: bytes=32 time=94ms TTL=128

Ping statistics for 192.168.2.2:
    Packets: Sent = 4, Received = 4, Lost = 0 (0% loss),
Approximate round trip times in milli-seconds:
    Minimum = 93ms, Maximum = 203ms, Average = 121ms
```

图 1.21　PC1 端使用 ping 命令示意图

7. 总　　结

本节主要介绍了交换机的 VLAN 技术。虚拟 VLAN 技术解决了交换机在进行局域网互连时无法限制广播的问题。VLAN 技术可以将一个物理局域网根据不同的原则划分成多个虚拟局域网，每个虚拟局域网形成一个广播域，从而实现了广播域划分切割，提高了网络运行的安全性及可靠性。

项目5 VLAN 中继配置

1. 工程目标

（1）了解 VLAN 通信原理。
（2）理解 VLAN 中继原理。
（3）掌握跨交换机实现 VLAN 间通信的方法。

2. 技术要点

VLAN 中继协议（VLAN Trunk Protocol，VTP）通过网络（CISCO 私有 DTP 帧）保持 VLAN 配置统一性，负责在 VTP 域内同步 VLAN 信息。可以添加、删除、更改局域网中 VLAN 信息，VTP 允许交换机共享并同步他们的 VLAN 信息，确保网络上 VLAN 信息一致。

VTP 消息只会通过干道传播。因此，需要在交换机之间设置干道以通过 VTP 来共享 VLAN 的信息。VTP 消息作为第二层组播帧传播，如果路由器将两台交换机分隔开，则路由器不会从它的一个接口转发 VTP 消息到另一个接口。

网络管理员在交换机上创建 VTP，并通过 VTP 来实现 VLAN 的管理。VTP 使用域来控制 VLAN，在每个 VTP 域中，可以将一台或多台交换机配置为 VTP 服务器。然后可以在 VTP 服务器上创建 VLAN，并将这些 VLAN 传送给域中的其他交换机。常见的 VTP 设置工作模式有 VTP 服务器、VTP 客户端和 VTP 透明网桥模式。

服务器模式（sever）：提供 VTP 消息，包括 VLAN ID 和名字信息，学习相同域名的 VTP 消息，转发相同域名的 VTP 消息，可以添加、删除、更改 VLAN，不同域名不学习、不转发。

客户机模式（client）：请求 VTP 消息，学习相同域名的 VTP 消息，转发相同域名的 VTP 消息，不能添加、删除、更改 VLAN，不同域名不学习、不转发。

透明模式（Transparent）：不提供 VTP 消息，不学习 VTP 消息，转发 VTP 消息，可以修改、删除、添加 VLAN，但只在本地生效。

3. 设备清单

交换机（3台）；网络连线（若干根）；测试 PC（6台）。

4. 工作场景

交换机基本配置工作场景如图1.22所示，主要设备有：3台思科二层交换机、6台 PC 终端。PC 终端分别通过网线连接到交换机上，3台二层交换机彼此之间通过网线相连。PC

终端通过网线对交换机进行基础配置和管理。

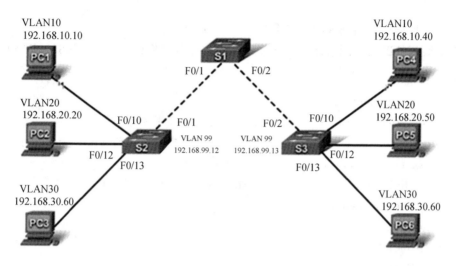

图 1.22　网络设备连接拓扑图

5. 地址规划

设备网络地址规划表如表 1.6 所示。

表 1.6　设备网络地址规划表

设备（主机）	接口	IP 地址	子网掩码	备注（网关、VLAN）
S1	VLAN 99	192.168.99.11	255.255.255.0	不适用
S2	VLAN 99	192.168.99.12	255.255.255.0	不适用
S3	VLAN 99	192.168.99.13	255.255.255.0	不适用
PC1	网卡	192.168.10.10	255.255.255.0	192.168.10.1
PC2	网卡	192.168.20.20	255.255.255.0	192.168.20.1
PC3	网卡	192.168.30.30	255.255.255.0	192.168.30.1
PC4	网卡	192.168.10.40	255.255.255.0	192.168.10.1
PC5	网卡	192.168.20.50	255.255.255.0	192.168.20.1
PC6	网卡	192.168.30.60	255.255.255.0	192.168.30.1
S2	F0/10			VLAN 10
S2	F0/12			VLAN 20
S2	F0/13			VLAN 30
S3	F0/10			VLAN 10
S3	F0/12			VLAN 20
S3	F0/13			VLAN 30

6. 工作过程

步骤一：根据拓扑图完成网络电缆连接。

根据实验拓扑图完成网络设备的连接，初始化配置交换机 S1、S2 和 S3。以 S1 为例：

Switch>enable
Switch#configure terminal
Switch(config)#hostname S1   //修改交换机名字为 S1
S1(config)#enable secret s1   //设置特权模式登录密码为 s1
S1(config)#no ip domain-lookup   //关闭域名解析功能
S1(config)#line console 0   //配置 console 口密码为 s1
S1(config-line)#password s1
S1(config-line)#login
S1(config-line)#line vty 0 10   //配置 vty 口密码为 s1
S1(config-line)#password s1
S1(config-line)#login

步骤二：配置主机 PC 上的以太网接口。

配置 PC1、PC2、PC3、PC4、PC5 和 PC6 的 IP 地址。以 PC1 为例，如图 1.23 所示。

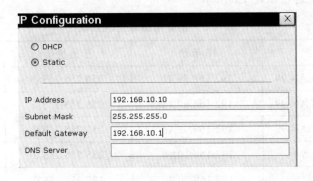

图 1.23　IP 地址配置图

步骤三：在交换机上配置 VTP。

为了使管理域更安全，域中每个交换机都需要配置域名和口令，并且域名和口令必须相同。

（1）配置 VTP 服务器、域名和密码。以 S1 为例：

S1(config)#vtp mode server   //将 S1 设置成服务器模式
S1(config)#vtp domain scitc   //设置 VTP 工作域名为 scitc
S1(config)#vtp password wl   //设置 VTP 密码为 wl

（2）配置 VTP 客户端。以 S2 为例：

S2(config)#vtp mode client   //将 S2 设置成客户端模式

S2(config)#vtp domain scitc　　　　//设置 VTP 工作域名为 scitc
S2(config)#vtp password cisco　　　//设置 VTP 密码为 cisco

（3）配置 VTP 透明网桥模式。以 S3 为例：

S3(config)#vtp mode transparent　　//将 S3 设置成透明模式
S2(config)#vtp domain scitc　　　　//设置 VTP 工作域名为 scitc
S2(config)#vtp password cisco　　　//设置 VTP 密码为 cisco

注：思科交换机默认 VTP 工作模式为 service 服务器模式，在配置前可以使用 show vtp status 查看配置情况，如图 1.24 所示。

```
s1#show vtp status
VTP Version                     : 2
Configuration Revision          : 4
Maximum VLANs supported locally : 255
Number of existing VLANs        : 9
VTP Operating Mode              : Server
VTP Domain Name                 :
VTP Pruning Mode                : Disabled
VTP V2 Mode                     : Disabled
VTP Traps Generation            : Disabled
MD5 digest                      : 0x7F 0xE2 0x89 0x85 0x7C 0x3D 0xB9 0xBD
Configuration last modified by 0.0.0.0 at 3-1-93 00:01:04
Local updater ID is 0.0.0.0 (no valid interface found)
```

图 1.24　VTP 查看配置图

步骤四：在 VTP 服务器上配置 VLAN，并使用 show vlan 命令查看 VLAN，如图 1.25 所示。

S1>enable
S1#vlan database
S1 (vlan)#vlan 99 name management　　//将 vlan 99 命名为 management
S1 (vlan)#vlan 10 name vlan10　　　　 //将 vlan 10 命名为 vlan10
S1 (vlan)#vlan 20 name vlan20　　　　 //将 vlan 20 命名为 vlan20
S1 (vlan)#vlan 30 name vlan30　　　　 //将 vlan 30 命名为 vlan30

```
Switch#show vlan

VLAN Name                         Status    Ports
---- -------------------------    --------- -------------------------------
1    default                      active    Fa0/1, Fa0/2, Fa0/3, Fa0/4
                                            Fa0/5, Fa0/6, Fa0/7, Fa0/8
                                            Fa0/9, Fa0/10, Fa0/11, Fa0/12
                                            Fa0/13, Fa0/14, Fa0/15, Fa0/16
                                            Fa0/17, Fa0/18, Fa0/19, Fa0/20
                                            Fa0/21, Fa0/22, Fa0/23, Fa0/24
                                            Gig1/1, Gig1/2
10   vlan10                       active
20   vlan20                       active
30   vlan30                       active
99   management                   active
1002 fddi-default                 act/unsup
1003 token-ring-default           act/unsup
1004 fddinet-default              act/unsup
```

图 1.25　查看 VLAN 基本信息

步骤五：配置交换机的中继端口。

S1(config)#interface range f0/1-2
S1(config-if-range)#switchport mode trunk   //将 f0/1-2 端口配置成 trunk 模式
S1(config-if-range)#switchport trunk native vlan 99
S1(config-if-range)#no shutdown
S1(config-if-range)#end

S2(config)# interface fa0/1
S2(config-if-range)#switchport mode trunk
S2(config-if-range)#switchport trunk native vlan 99
S2(config-if-range)#no shutdown
S2(config-if-range)#end

S3(config)# interface fa0/2
S3(config-if-range)#switchport mode trunk
S3(config-if-range)#switchport trunk native vlan 99
S3(config-if-range)#no shutdown
S3(config-if-range)#end

步骤六：S3 配置 VLAN。

S3>enable
S3#vlan database
S3 (vlan)#vlan 99 name management   //将 vlan 99 命名为 management
S3 (vlan)#vlan 10 name vlan10   //将 vlan 10 命名为 vlan10
S3 (vlan)#vlan 20 name vlan20   //将 vlan 20 命名为 vlan20
S3 (vlan)#vlan 30 name vlan30   //将 vlan 30 命名为 vlan30

步骤七：设置 IP 地址。

S1(config)#interface vlan 99   //设置 vlan 99 IP 地址
S1(config-if)#ip address 192.168.99.11 255.255.255.0
S1(config-if)#no shutdown
S2(config)#interface vlan 99   //设置 vlan 99 IP 地址
S2(config-if)#ip address 192.168.99.12 255.255.255.0
S2(config-if)#no shutdown
S3(config)#interface vlan 99   //设置 vlan 99 IP 地址
S3(config-if)#ip address 192.168.99.13 255.255.255.0
S3(config-if)#no shutdown

步骤八：将交换机端口分配给 VLAN。
S2#configure terminal
S2(config)#interface f0/10
S2(config-if)#switchport access vlan 10
S2(config-if)#no shutdown
S2#configure terminal
S2(config)#interface f0/12
S2(config-if)#switchport access vlan 20
S2(config-if)#no shutdown
S2#configure terminal
S2(config)#interface f0/13
S2(config-if)#switchport access vlan 30
S2(config-if)#no shutdown

S3#configure terminal
S3(config)#interface f0/10
S3(config-if)#switchport access vlan 10
S3(config-if)#no shutdown
S3#configure terminal
S3(config)#interface f0/12
S3(config-if)#switchport access vlan 20
S3(config-if)#no shutdown
S3#configure terminal
S3(config)#interface f0/13
S3(config-if)#switchport access vlan 30
S3(config-if)#no shutdown

步骤九：使用 show vlan 分别在 S2 和 S3 上查看 VLAN 信息。

7. 总　　结

本节主要介绍了 VLAN 中的中继技术，通过 VLAN 中继实现了 VLAN 的跨交换机通信，使 VLAN 技术的可使用性得到了提高，满足了人们日常工作的需要。

项目 6 链路聚合配置

1. 工程目标

（1）EtherChannel 的工作原理。
（2）EtherChannel 的配置。

2. 技术要点

1）EtherChannel 技术目标和用途

EtherChannel 是 Cisco 公司开发的技术，通过将两台交换机之间多条以太链路捆绑在一起组成一条逻辑链路，增加交换机之间数据传输带宽（Cisco 交换机最多允许绑定 8 个交换机端口到一个逻辑链路中），EtherChannel 可采用负载均衡技术在多条链路上均衡分配流量。启用了 EtherChannel 技术的服务器也可以和交换机之间采用 EtherChannel 技术。

2）EtherChannel 协议类型

为了描述清楚，定义参加链路聚合的交换机端口为物理端口，链路聚合后的逻辑端口为逻辑端口。

EtherChannel 有两个管理协议：一个是 PAgP（端口聚合协议），它是 Cisco 私有协议；另一个是 LACP 协议，它是 IEEE 的标准协议。这些协议可以确保在创建好 EtherChannel 时，在同一个交换机中捆绑到同一逻辑链路的物理端口的配置是相同的；在配置 EtherChannel 时同一逻辑链路包含的各个物理端口的速度、双工模式、VLAN 信息的设置必须完全一致。EtherChannel 配置完成后对逻辑端口的配置修改会影响到每个物理端口。

Cisco 网络环境（只有 Cisco 网络设备）中可以使用 PAgP 协议或 LACP 协议；混合网络环境（既有 Cisco 网络设备又有其他厂商网络设备）只能够使用 LACP 协议。

PAgP 模式下两端逻辑端口的模式需要兼容，否则无法建立 EtherChannel。尤其要注意两边都是 Auto 模式时是无法协商建立 EtherChannel 的。两端在不同模式下协商成功与否具体情况如图 1.26 所示。

LACP 模式和 PAgP 模式类似，协商过程需要两端逻辑端口兼容，两端在不同模式下协商成功与否具体况如图 1.27 所示。

3）EtherChannel 负载均衡

EtherChannel 将多个物理链路逻辑捆绑在一起，提高了交换机之间的通信速度，同时它支持用户手工配置物理链路的负载分担，可选的配置有基于源或者目的 MAC，基于源或目标的 IP 地址、源或者目标的 TCP/UDP 端口、源和目标端的 TCP/IP 端口。

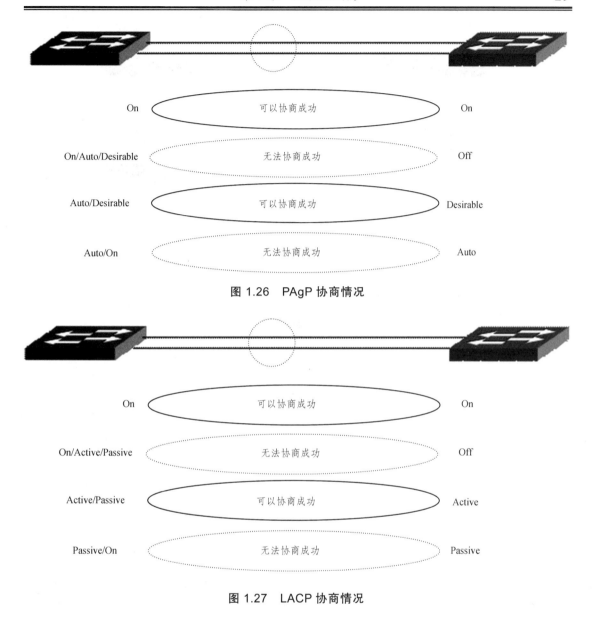

图 1.26　PAgP 协商情况

图 1.27　LACP 协商情况

3. 设备清单

二层交换机（2 台）；网络连线（若干根）；测试 PC（2 台）。

4. 工作场景

为整合教学资源，某信息大学把原电子工程系、信息工程系及相关专业在院系改建中合并成电子信息学院，并将原来各自分隔的网络连成一体，升级学院主干网络。此外为保证网络的稳定性，在主干网络改造的过程中，增加网络冗余。如图 1.28 所示，2 台二层交换机之间需要配置链路聚合，PC1 和 PC2 作为 2 个客户端，分别接入 switch1 和 switch2。

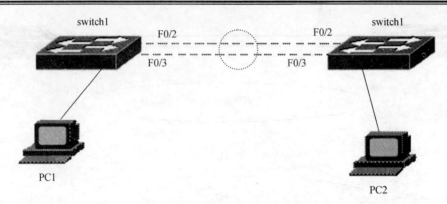

图 1.28 二层交换机链路聚合拓扑

5. 地址规划

二层交换机链路地址规划表如表 1.7 所示。

表 1.7 二层交换机链路地址规划表

设备名称	设备及端口的配置地址	备注
switch1	Fa0/1	连接 PC1
	Fa0/2-3	Channel-group 2，连接 switch2 对应端口
switch2	Fa0/1	连接 PC2
	Fa0/2-3	Channel-group 5，连接 switch1 对应端口
PC1	172.16.1.1/24	vlan1
PC2	172.16.1.2/24	vlan1

6. 工作过程

步骤一：连接设备。如图 1.28 所示使用网络线在工作现场连接好设备。建议先规划好拓扑图及聚合端口，不要将两个交换机之间的线缆连接起来，容易引起环路。

步骤二：配置交换机基本信息。

步骤三：完成链路聚合配置

（1）配置 switch1。

Switch>enable
switch#configure terminal
Enter configuration commands, one per line. End with CNTL/Z.
Switch(config)#hostname switch1
switch1(config)#interface fastEthernet 0/2
switch1(config-if)#channel-group 2 mode active
switch1(config-if)#exit

switch1(config)#interface fastEthernet 0/3
switch1(config-if)#channel-group 2 mode active
switch1(config-if)#exit
switch1(config)#interface port-channel 2
switch1(config-if)#switchport mode trunk
switch1(config-if)#switchport trunk allowed vlan   1,2,3,99

（2）配置 switch2。

Switch>enable
switch#configure terminal
Enter configuration commands, one per line. End with CNTL/Z.
Switch(config)#hostname switch2
Switch2(config)#interface fastEthernet 0/2
Switch2(config-if)#channel-group 5 mode on
Switch2(config-if)#exit
Switch2(config)#interface fastEthernet 0/3
Switch2(config-if)#channel-group 5 mode on
Switch2(config-if)#exit
Switch2(config)#interface port-channel 5
Switch2(config-if)#switchport mode trunk
Switch2(config-if)#switchport trunk allowed vlan   1,2,3,99

（3）配置测试计算机地址，测试网络连通性。
① 配置 PC1 的地址为 172.16.1.1/24，配置 PC2 的地址为 172.16.1.2/24。
② PC 上使用 ping 命令测试网络连通性。
（4）验证 EtherChannel 配置。

switch1#show etherchannel port-channel
switch1#show etherchannel summary

7．总　　结

为了提高网络带宽，网络管理员可以使用 EtherChannel 技术将交换机之间独立的链路捆绑成一个逻辑链路。这个技术可用在二层交换机之间、三层交换机之间以及交换机与开启了 EtherChannel 支持功能的服务器之间。EtherChannel 有两种协议可以使用：一种是 Cisco 私有的协议 PAgP 协议；另一种是国际标准协议 LCAP。使用 EtherChannel 的难点在于逻辑端口的模式配置，只有两端的逻辑端口相互兼容才能够建立 EtherChannel 隧道，管理员可以根据需要配置链路上的流量负荷。

项目7 生成树协议（STP）配置

1. 工程目标

（1）了解 STP 原理及用途。
（2）掌握 STP 的配置。

2. 技术要点

STP（Spanning Tree Protocol）协议主要用于在冗余链路中，通过算法实现没有环路的网络。生成树协议是利用 SPA 算法，在存在交换环路的网络中生成一个没有环路的树形网络。运用该算法将交换网络的冗余备份链路从逻辑上断开，当主链路出现故障时，能够自动切换到备份链路，保证数据的正常转发。

生成树协议版本：STP、RSTP（快速生成树）、MSTP（多生成树协议）。

生成树协议运行生成树算法（STA）：

（1）选择根网桥（根据交换机的优先级来确定）。
（2）选择根端口（根网桥的最低路径成本较低、直连的网桥 ID 最小、端口 ID 最小）。
（3）选择指定端口（根路径成本较低、所在交换机网桥 ID 值最小、端口 ID 值小）。

3. 设备清单

二层交换机（3台）；网络连线（若干根）。

4. 工作场景

智翔学院、信息学院和联想学院为了保证网络稳定性，在信息学院与联想学院间实现冗余，三个学院网络连接情况如图 1.29 所示，设置智翔学院为根网桥。

5. 工作过程

步骤一：连接设备。如图 1.29 所示，使用网络线在工作现场连接好设备。
步骤二：配置每个接口模式为 Trunk，在每个交换机上配置一个 VLAN2，查看 STP 状态，修改 S 的优先级，然后查看 STP 的状态，比较每个端口发生的变化，理解 STP 原理。

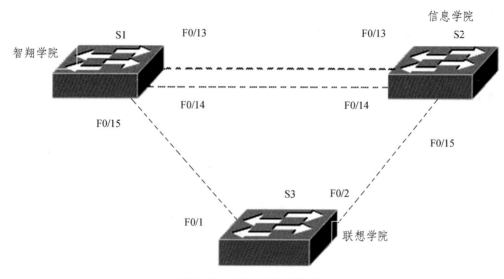

图 1.29 STP 实验拓扑图

（1）S1 配置。

Switch>en
Switch#configure terminal
Switch(config)#hostname S1
S1(config)#interface f0/13
S1(config-if)#switchport mode trunk
S1(config-if)#exit
S1(config)#int f0/14
S1(config-if)#switchport mode trunk
S1(config-if)#exit
S1(config)#int f0/15
S1(config-if)#switchport mode trunk
S1(config-if)#end
S1(config)#vtp domain STP-TEST
S1(config)#vlan 2
S1(config-vlan)#exit

（2）S2 配置。

Switch>en
Switch#configure terminal
Switch(config)#hostname S2
S2(config)#int f0/13
S2(config-if)#switchport mode trunk
S2(config-if)#exit

S2(config)#interface f0/14
S2(config-if)#switchport mode trunk
S2(config-if)#exit
S2(config)#interface f0/15
S2(config-if)#switchport mode trunk
S2(config-if)#exit

（3）S3 配置。

S3(config)#int f0/1
S3(config-if)#switchport mode trunk
S3(config-if)#exit
S3(config)#int f0/2
S3(config-if)#switchport mode trunk
S3(config-if)#end

（4）在 S1 上查看 STP 状态。

S1#show spanning-tree
VLAN0001
Spanning tree enabled protocol ieee
Root ID Priority 32769
Address 0001.4398.043A
This bridge is the root
Hello Time 2 sec Max Age 20 sec Forward Delay 15 sec

Bridge ID Priority 32769 (priority 32768 sys-id-ext 1)
Address 0001.4398.043A
Hello Time 2 sec Max Age 20 sec Forward Delay 15 sec
Aging Time 20

Interface Role Sts Cost Prio.Nbr Type
---------------- ---- --- --------- -------- ------------------------
Fa0/13 Desg FWD 19 128.13 P2p
Fa0/14 Desg FWD 19 128.14 P2p
Fa0/15 Desg FWD 19 128.15 P2p

VLAN0002
Spanning tree enabled protocol ieee
Root ID Priority 32770
Address 0001.4398.043A

This bridge is the root
Hello Time 2 sec Max Age 20 sec Forward Delay 15 sec

Bridge ID Priority 32770 (priority 32768 sys-id-ext 2)
Address 0001.4398.043A
Hello Time 2 sec Max Age 20 sec Forward Delay 15 sec
Aging Time 20

Interface Role Sts Cost Prio.Nbr Type
---------------- ---- --- --------- -------- -----------------------
Fa0/13 Desg FWD 19 128.13 P2p
Fa0/14 Desg FWD 19 128.14 P2p
Fa0/15 Desg FWD 19 128.15 P2p

（5）在 S2 上查看 STP 状态。

S2#show spanning-tree
VLAN0001
Spanning tree enabled protocol ieee
Root ID Priority 32769
Address 0001.4398.043A
Cost 19
Port 13(FastEthernet0/13)
Hello Time 2 sec Max Age 20 sec Forward Delay 15 sec

Bridge ID Priority 32769 (priority 32768 sys-id-ext 1)
Address 0001.C733.0D73
Hello Time 2 sec Max Age 20 sec Forward Delay 15 sec
Aging Time 20

Interface Role Sts Cost Prio.Nbr Type
---------------- ---- --- --------- -------- -----------------------
Fa0/14 Altn BLK 19 128.14 P2p
Fa0/13 Root FWD 19 128.13 P2p
Fa0/15 Desg FWD 19 128.15 P2p

VLAN0002
Spanning tree enabled protocol ieee
Root ID Priority 32770
Address 0001.4398.043A

Cost 19
Port 13(FastEthernet0/13)
Hello Time 2 sec Max Age 20 sec Forward Delay 15 sec

Bridge ID Priority 32770 (priority 32768 sys-id-ext 2)
Address 0001.C733.0D73
Hello Time 2 sec Max Age 20 sec Forward Delay 15 sec
Aging Time 20

Interface Role Sts Cost Prio.Nbr Type
---------------- ---- --- --------- -------- ----------------------
Fa0/14 Altn BLK 19 128.14 P2p
Fa0/13 Root LRN 19 128.13 P2p
Fa0/15 Desg LRN 19 128.15 P2p

（6）S3 上查看 STP 状态。

S3#show spanning-tree
VLAN0001
Spanning tree enabled protocol ieee
Root ID Priority 32769
Address 0001.4398.043A
Cost 19
Port 1(FastEthernet0/1)
Hello Time 2 sec Max Age 20 sec Forward Delay 15 sec

Bridge ID Priority 32769 (priority 32768 sys-id-ext 1)
Address 0009.7C7A.CCDE
Hello Time 2 sec Max Age 20 sec Forward Delay 15 sec
Aging Time 20

Interface Role Sts Cost Prio.Nbr Type
---------------- ---- --- --------- -------- ----------------------
Fa0/2 Altn BLK 19 128.2 P2p
Fa0/1 Root FWD 19 128.1 P2p

VLAN0002
Spanning tree enabled protocol ieee
Root ID Priority 32770
Address 0001.4398.043A

Cost 19
Port 1(FastEthernet0/1)
Hello Time 2 sec Max Age 20 sec Forward Delay 15 sec

Bridge ID Priority 32770 (priority 32768 sys-id-ext 2)
Address 0009.7C7A.CCDE
Hello Time 2 sec Max Age 20 sec Forward Delay 15 sec
Aging Time 20

Interface Role Sts Cost Prio.Nbr Type
---------------- ---- --- --------- -------- ------------------------
Fa0/2 Altn BLK 19 128.2 P2p
Fa0/1 Root FWD 19 128.1 P2p

S1(config)#spanning-tree vlan 1 priority 4096
S1(config)#end
S1#show spanning-tree
VLAN0001
Spanning tree enabled protocol ieee
Root ID Priority 4097
Address 0001.4398.043A
This bridge is the root
Hello Time 2 sec Max Age 20 sec Forward Delay 15 sec

Bridge ID Priority 4097 (priority 4096 sys-id-ext 1)
Address 0001.4398.043A
Hello Time 2 sec Max Age 20 sec Forward Delay 15 sec
Aging Time 20

Interface Role Sts Cost Prio.Nbr Type
---------------- ---- --- --------- -------- ------------------------
Fa0/13 Desg FWD 19 128.13 P2p
Fa0/14 Desg FWD 19 128.14 P2p
Fa0/15 Desg FWD 19 128.15 P2p

VLAN0002
Spanning tree enabled protocol ieee
Root ID Priority 32770
Address 0001.4398.043A

This bridge is the root
Hello Time 2 sec Max Age 20 sec Forward Delay 15 sec

Bridge ID Priority 32770 (priority 32768 sys-id-ext 2)
Address 0001.4398.043A
Hello Time 2 sec Max Age 20 sec Forward Delay 15 sec
Aging Time 20

Interface Role Sts Cost Prio.Nbr Type
---------------- ---- --- --------- -------- -----------------------
Fa0/13 Desg FWD 19 128.13 P2p
Fa0/14 Desg FWD 19 128.14 P2p
Fa0/15 Desg FWD 19 128.15 P2p

（7）在 S1 上修改优先级。

S1#configure terminal
S1(config)#spanning-tree vlan 1 priority 4096

（8）S1 上查看 STP 状态。

Switch#show spanning-tree
VLAN0001
Spanning tree enabled protocol ieee
Root ID Priority 4097
Address 0001.4398.043A
Cost 19
Port 13(FastEthernet0/13)
Hello Time 2 sec Max Age 20 sec Forward Delay 15 sec

Bridge ID Priority 32769 (priority 32768 sys-id-ext 1)
Address 0001.C733.0D73
Hello Time 2 sec Max Age 20 sec Forward Delay 15 sec
Aging Time 20

Interface Role Sts Cost Prio.Nbr Type
---------------- ---- --- --------- -------- ----------------------------------
Fa0/14 Altn BLK 19 128.14 P2p
Fa0/13 Root FWD 19 128.13 P2p
Fa0/15 Desg FWD 19 128.15 P2p

VLAN0002
Spanning tree enabled protocol ieee
Root ID Priority 32770
    Address 0001.4398.043A
    Cost 19
    Port 13(FastEthernet0/13)
    Hello Time 2 sec Max Age 20 sec Forward Delay 15 sec

Bridge ID Priority 32770 (priority 32768 sys-id-ext 2)
    Address 0001.C733.0D73
    Hello Time 2 sec Max Age 20 sec Forward Delay 15 sec
    Aging Time 20

Interface Role Sts Cost Prio.Nbr Type
---------------- ---- --- --------- -------- --------------------
Fa0/14 Altn BLK 19 128.14 P2p
Fa0/13 Root FWD 19 128.13 P2p
Fa0/15 Desg FWD 19 128.15 P2p

（9）S2 上查看 STP 状态。

S2#show spanning-tree
VLAN0001
Spanning tree enabled protocol ieee
Root ID Priority 32769
    Address 0001.4398.043A
    Cost 19
    Port 1(FastEthernet0/1)
    Hello Time 2 sec Max Age 20 sec Forward Delay 15 sec

Bridge ID Priority 32769 (priority 32768 sys-id-ext 1)
    Address 0009.7C7A.CCDE
    Hello Time 2 sec Max Age 20 sec Forward Delay 15 sec
    Aging Time 20

Interface Role Sts Cost Prio.Nbr Type
---------------- ---- --- --------- -------- --------------------------------
Fa0/2 Altn BLK 19 128.2 P2p
Fa0/1 Root FWD 19 128.1 P2p

VLAN0002
Spanning tree enabled protocol ieee
Root ID Priority 32770
Address 0001.4398.043A
Cost 19
Port 1(FastEthernet0/1)
Hello Time 2 sec Max Age 20 sec Forward Delay 15 sec

Bridge ID Priority 32770 (priority 32768 sys-id-ext 2)
Address 0009.7C7A.CCDE
Hello Time 2 sec Max Age 20 sec Forward Delay 15 sec
Aging Time 20

Interface Role Sts Cost Prio.Nbr Type
---------------- ---- --- --------- -------- --------------------------------
Fa0/2 Altn BLK 19 128.2 P2p
Fa0/1 Root FWD 19 128.1 P2p

（10）S3 上查看 STP 状态。

S3#show spanning-tree
VLAN0001
Spanning tree enabled protocol ieee
Root ID Priority 4097
Address 0001.4398.043A
Cost 19
Port 1(FastEthernet0/1)
Hello Time 2 sec Max Age 20 sec Forward Delay 15 sec

Bridge ID Priority 32769 (priority 32768 sys-id-ext 1)
Address 0009.7C7A.CCDE
Hello Time 2 sec Max Age 20 sec Forward Delay 15 sec
Aging Time 20

Interface Role Sts Cost Prio.Nbr Type
---------------- ---- --- --------- -------- --------------------------------
Fa0/2 Altn BLK 19 128.2 P2p
Fa0/1 Root FWD 19 128.1 P2p

VLAN0002
Spanning tree enabled protocol ieee
Root ID Priority 32770
Address 0001.4398.043A
Cost 19
Port 1(FastEthernet0/1)
Hello Time 2 sec Max Age 20 sec Forward Delay 15 sec

Bridge ID Priority 32770 (priority 32768 sys-id-ext 2)
Address 0009.7C7A.CCDE
Hello Time 2 sec Max Age 20 sec Forward Delay 15 sec
Aging Time 20

Interface Role Sts Cost Prio.Nbr Type
---------------- ---- --- --------- -------- --------------------------------
Fa0/2 Altn BLK 19 128.2 P2p
Fa0/1 Root FWD 19 128.1 P2p

6. 总 结

STP 协议的目的是解决网络中冗余链路带来的环路问题。在一般的网络中系统会自动启用 STP 协议，所以我们需要掌握 STP 协议的原理，根据网络环境熟练控制 STP 协议。

项目8 快速生成树（RSTP）配置

1. 工程目标

（1）理解快速生成树协议 RSTP 的工作原理。
（2）掌握快速生成树 RSTP 的配置方法。

2. 技术要点

RSTP（IEEE802.1w）是一种快速生成树协议，是在 STP 技术中融入了减少收敛时间的措施后形成的，能够达到相当快的收敛速度，有的甚至只需要几百毫秒。

快速生成树协议在生成树协议的基础上增加了两种端口角色：替换端口和备份端口，分别作为根端口和指定端口的冗余端口。当根端口或指定端口出现故障时，冗余端口不需要经过 50 s 的收敛时间，可以直接切换到替换端口或备份端口，从而实现 RSTP 协议小于 1 s 的快速收敛。

RSTP 的特征有：
（1）集成了 IEEE802.1d 的很多增强技术，这些增强功能不需要额外的配置。
（2）RSTP 使用与 IEEE802.1d 相同的 BPDU 格式。
（3）RSTP 能够主动确认端口是否能够安全转换到转发状态，而不需要依靠任何计时器来做出判断。

3. 设备清单

路由器（2台）；网络连线（若干根）；测试 PC（2台）。

4. 工作场景

如图 1.30 所示，2 台二层交换机之间连接了两条物理链路，在 SwitchA 和 SwitchB 上配置快速生成树。

5. 工作过程

步骤一：连接设备。如图 1.30 所示使用网络线在工作现场连接好设备。
步骤二：交换机基本 VLAN 创建，接口配置，生成树配置，查看生成树信息。
步骤三：修改 SwitchB 中 vlan10 的优先级，对比生成树的变化，掌握快速生成树的原理。

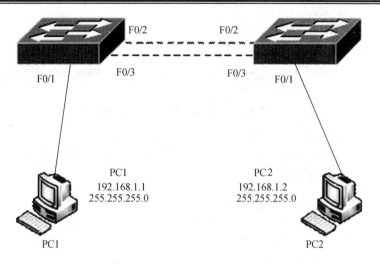

图 1.30　RSTP 网络拓扑

（1）配置 SwitchB。

Switch>

Switch>en

Switch#configure terminal

Switch(config)#hostname SwitchB

SwitchB(config)#vlan 10

SwitchB(config-vlan)#exit

SwitchB(config)#int f0/1

SwitchB(config-if)#switchport mode access

SwitchB(config-if)#switchport access vlan 10

SwitchB(config)#interface f0/2

SwitchB(config-if)#switchport mode access

SwitchB(config-if)#switchport mode trunk

SwitchB(config-if)#exit

SwitchB(config)#int f0/3

SwitchB(config-if)#switchport mode access

SwitchB(config-if)#switchport mode trunk

SwitchB(config-if)#exit

SwitchB(config)#spanning-tree mode rapid-pvst

SwitchB(config)#end

SwitchB#show spanning-tree

VLAN0001

　Spanning tree enabled protocol rstp

　Root ID　　　Priority　　　32769

　　　　　　　Address　　　　0001.6484.12A6

         Cost   19
         Port   2(FastEthernet0/2)
         Hello Time   2 sec   Max Age 20 sec   Forward Delay 15 sec

  Bridge ID   Priority   32769   (priority 32768 sys-id-ext 1)
              Address   0001.C9AA.38E7
              Hello Time   2 sec   Max Age 20 sec   Forward Delay 15 sec
              Aging Time   20

Interface   Role Sts Cost   Prio.Nbr Type
---------------- ---- --- --------- -------- --------------------------------
Fa0/2   Root FWD 19   128.2   P2p
Fa0/3   Altn BLK 19   128.3   P2p

VLAN0010
  Spanning tree enabled protocol rstp
  Root ID   Priority   32778
             Address   0001.6484.12A6
             Cost   19
             Port   2(FastEthernet0/2)
             Hello Time   2 sec   Max Age 20 sec   Forward Delay 15 sec

  Bridge ID Priority   32778   (priority 32768 sys-id-ext 10)
             Address   0001.C9AA.38E7
             Hello Time   2 sec   Max Age 20 sec   Forward Delay 15 sec
             Aging Time   20

Interface   Role Sts Cost   Prio.Nbr Type
---------------- ---- --- --------- -------- --------------------------------
Fa0/2   Root FWD 19   128.2   P2p
Fa0/3   Altn BLK 19   128.3   P2p

（2）交换机 B 的配置。
Switch>en
Switch#configure terminal
Enter configuration commands, one per line.   End with CNTL/Z.
Switch(config)#hos
Switch(config)#hostname SwitchB
SwitchB(config)#vlan 10
SwitchB(config-vlan)#exit

```
SwitchB(config)#int f0/1
SwitchB(config-if)#switchport mode access
SwitchB(config-if)#switchport access vlan 10
SwitchB(config-if)#exit
SwitchB(config)#int f0/2
SwitchB(config-if)#switchport mode access
SwitchB(config-if)#switchport mode trunk
SwitchB(config-if)#exit
SwitchB(config)#int f0/3
SwitchB(config-if)#switchport mode access
SwitchB(config-if)#switchport mode trunk
SwitchB(config-if)#exit
SwitchB(config)#spanning-tree mode rapid-pvst
SwitchB(config)#end
SwitchB#show spanning-tree
VLAN0001
  Spanning tree enabled protocol rstp
  Root ID    Priority    32769
             Address     0001.6484.12A6
             This bridge is the root
             Hello Time  2 sec  Max Age 20 sec  Forward Delay 15 sec

  Bridge ID  Priority    32769  (priority 32768 sys-id-ext 1)
             Address     0001.6484.12A6
             Hello Time  2 sec  Max Age 20 sec  Forward Delay 15 sec
             Aging Time  20

Interface        Role Sts Cost      Prio.Nbr Type
---------------- ---- --- --------- -------- --------------------------------
Fa0/2            Desg FWD 19         128.2    P2p
Fa0/3            Desg FWD 19         128.3    P2p


VLAN0010
  Spanning tree enabled protocol rstp
  Root ID    Priority    32778
             Address     0001.6484.12A6
             This bridge is the root
             Hello Time  2 sec  Max Age 20 sec  Forward Delay 15 sec
```

       Bridge ID   Priority   32778   (priority 32768 sys-id-ext 10)
                    Address   0001.6484.12A6
                    Hello Time   2 sec   Max Age 20 sec   Forward Delay 15 sec
                    Aging Time   20

Interface   Role Sts Cost   Prio.Nbr Type
---------------- ---- --- --------- -------- --------------------------------
Fa0/2   Desg FWD 19   128.2   P2p
Fa0/3   Desg FWD 19   128.3   P2p

（3）修改 SwitchA 中 VLAN 1 的优先级，查看生成树的变化。

SwitchA#configure terminal
SwitchA(config)#spanning-tree vlan 1 priority 4096
SwitchA(config)#end
SwitchA#show spanning-tree
VLAN0001
  Spanning tree enabled protocol rstp
  Root ID   Priority   4097
               Address   0001.C9AA.38E7
               This bridge is the root
               Hello Time   2 sec   Max Age 20 sec   Forward Delay 15 sec

  Bridge ID   Priority   4097   (priority 4096 sys-id-ext 1)
               Address   0001.C9AA.38E7
               Hello Time   2 sec   Max Age 20 sec   Forward Delay 15 sec
               Aging Time   20

Interface   Role Sts Cost   Prio.Nbr Type
---------------- ---- --- --------- -------- --------------------------------
Fa0/2   Desg FWD 19   128.2   P2p
Fa0/3   Desg FWD 19   128.3   P2p

VLAN0010
  Spanning tree enabled protocol rstp
  Root ID   Priority   32778
               Address   0001.6484.12A6
               Cost   19
               Port   2(FastEthernet0/2)
               Hello Time   2 sec   Max Age 20 sec   Forward Delay 15 sec

```
    Bridge ID   Priority     32778   (priority 32768 sys-id-ext 10)
                Address      0001.C9AA.38E7
                Hello Time   2 sec   Max Age 20 sec   Forward Delay 15 sec
                Aging Time   20

Interface         Role Sts Cost      Prio.Nbr Type
---------------- ---- --- --------- -------- --------------------------------
Fa0/2             Root FWD 19          128.2    P2p
Fa0/3             Altn BLK 19          128.3    P2p
```

（4）查看生成树的情况。

```
SwitchB#show spanning-tree
VLAN0001
  Spanning tree enabled protocol rstp
  Root ID     Priority    4097
              Address     0001.C9AA.38E7
              Cost        19
              Port        2(FastEthernet0/2)
              Hello Time  2 sec   Max Age 20 sec   Forward Delay 15 sec

  Bridge ID   Priority    32769   (priority 32768 sys-id-ext 1)
              Address     0001.6484.12A6
              Hello Time  2 sec   Max Age 20 sec   Forward Delay 15 sec
              Aging Time  20

Interface         Role Sts Cost      Prio.Nbr Type
---------------- ---- --- --------- -------- --------------------------------
Fa0/2             Root FWD 19          128.2    P2p
Fa0/3             Altn BLK 19          128.3    P2p

VLAN0010
  Spanning tree enabled protocol rstp
  Root ID     Priority    32778
              Address     0001.6484.12A6
              This bridge is the root
              Hello Time  2 sec   Max Age 20 sec   Forward Delay 15 sec

  Bridge ID   Priority    32778   (priority 32768 sys-id-ext 10)
```

```
             Address         0001.6484.12A6
             Hello Time   2 sec   Max Age 20 sec   Forward Delay 15 sec
             Aging Time   20

Interface        Role Sts Cost        Prio.Nbr Type
---------------- ---- --- ---------   --------------------------------
Fa0/2            Desg FWD 19          128.2    P2p
Fa0/3            Desg FWD 19          128.3    P2p
```

6. 总　结

RSTP是生成树的一种，其特点是能够快速地收敛（收敛指的是交换机达到生成树最后的结果），配置内容不是很多，重点是掌握生成树的原理，理解其工作过程。

项目 9 VLAN 间路由配置

1. 工程目标

掌握三层交换机 VLAN 间通信的原理及配置技术。

2. 技术要点

由于 VLAN 能够将流量隔离到确定的广播域和子网中，不同的 VLAN 中的设备不能直接的相互通信，所以只有通过三层设备才能让不同 VLAN 中的设备之间进行通信，这也被称为 VLAN 间的路由。三层设备有路由器和三层交换机，它们都能够实现 VLAN 间路由。如图 1.31 所示。

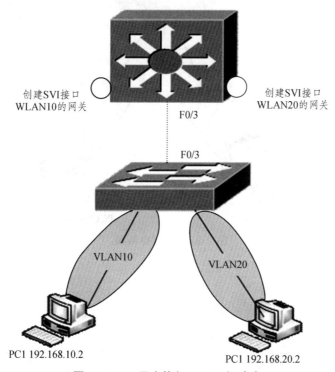

图 1.31 三层交换机 VLAN 间路由

三层交换机 VLAN 间通信的原理是：在三层交换机上创建 VLAN 即 SVI 接口，每个 SVI 接口对应一个 VLAN 的网关，开启三层交换机的路由功能，可以实现不同 VLAN 间的数据信息的路由。

3. 设备清单

三层交换机（1 台）；二层交换机（1 台）；网络连线（若干根）；测试 PC（2 台）。

4. 工作场景

如图 1.32 所示,模拟局域网中两个用户、接入层及汇聚层的连接方式,采用三层交换实现 VLAN 间的路由。

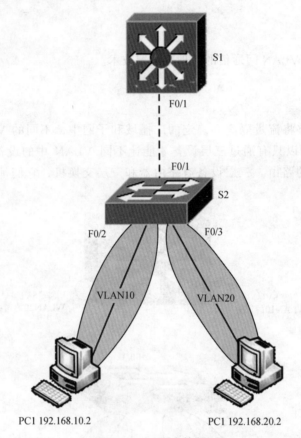

图 1.32 三层交换机

5. 地址规划

三层交换机地址规划如表 1.8 所示。

表 1.8 地址规划

设备名称	设备及端口的配置地址		备注
S1	Fa0/1	配置为 Trunk 模式	
	VLAN 10	192.168.10.1/24	
	VLAN 20	192.168.20.1/24	
S2	F0/1	配置为 trunk 模式	
	F0/2	连接 PC1	
	F0/3	连接 PC2	
PC1	192.168.10.2/24		网关:192.168.10.1
PC2	192.168.20.2/24		网关:192.168.20.1

6. 工作过程

步骤一：根据图 1.32 连接设备。

步骤二：在交换机上配置 VLAN，将 PC1 和 PC2 连接的接口分配到对应的 VLAN 中，配置交换机连接的接口为 trunk 模式。

步骤三：三层交换机 S1 上配置 SVI 接口，分配相应的 IP 地址。

步骤四：配置 PC1、PC2 的地址及网关。

（1）配置 S2。

```
Switch>en
Switch#configure terminal
Switch(config)#vlan 10
Switch(config-vlan)#vlan 20
Switch(config-vlan)#exit
Switch(config)#int f0/2
Switch(config-if)#switchport mode access
Switch(config-if)#switchport access vlan 10
Switch(config-if)#exit
Switch(config)#int f0/3
Switch(config-if)#switchport mode access
Switch(config-if)#switchport access vlan 20
Switch(config-if)#exit
Switch(config)#int f0/1
Switch(config-if)#switchport mode trunk
```

（2）配置 S1。

```
Switch>en
Switch#configure terminal
Switch(config)#int f0/1
Switch(config-if)#switchport mode access
Switch(config-if)#switchport mode trunk
Switch(config-if)#exit
Switch(config)#vlan 10
Switch(config-vlan)#vlan 20
Switch(config-vlan)#exit
Switch(config)#int vlan 10
Switch(config-if)#ip add 192.168.10.1 255.255.255.0
Switch(config-if)#exit
Switch(config)#int vlan 20
Switch(config-if)#ip add 192.168.20.1 255.255.255.0
```

Switch(config-if)#exit
Switch(config)#ip routing

（3）配置主机 IP 地址和网关，测试 PC1 能否 ping 通 PC2。

7. 总　　结

为了实现不同 VLAN 间的计算机能够相互通信，需要三层设备在不同网段之间路由数据，路由器采用为每个网段配置子接口，作为 VLAN 的网关，路由器将为各个 VLAN 路由数据。三层交换机首先开启路由功能，然后 SVI 接口作为每个 VLAN 的网关，由三层交换机提供路由功能。容易出现的问题是三层交换机中 SVI 接口没有 UP，解决的办法是先创建 VLAN，再用命令 interface vlan x，才能正常开启 SVI 接口。

第 2 章 路由实践篇

项目 10 路由器基本配置

1. 工程目标

（1）根据拓扑图进行网络布线。
（2）认识路由器及其基本配置命令。
（3）配置并激活以太网接口。

2. 技术要点

路由器是一种典型的网络层设备，它在两个局域网之间接帧传输数据。在 OSI/RM 之中被称之为中介系统，完成网络层中继或第三层中继的任务。路由器负责在两个局域网的网络层间接帧传输数据，转发帧时需要改变帧中的地址。

路由器可连接多个网络，具有多个接口，每个接口属于不同的 IP 网络。当路由器从某个接口收到 IP 数据包时，它会确定使用哪个接口来将该数据包转发到目的地。路由器用于转发数据包的接口可以位于数据包的最终目的网络（即具有该数据包目的 IP 地址的网络），也可以位于连接到其他路由器的网络（用于送达目的网络）。

路由器连接的每个网络通常需要单独的接口。这些接口用于连接局域网（LAN）和广域网（WAN）。LAN 通常称为以太网，其中包含各种设备，例如：PC、打印机和服务器。WAN 用于连接分布在广阔地域中的网络。例如：WAN 连接通常用于将 LAN 连接到 Internet 服务提供商（ISP）网络。

3. 设备清单

路由器（1 台）；交换机（1 台）；网络连线（若干根）；测试 PC（2 台）。

4. 工作场景

某小型企业在科技大厦 3 楼租了 2 间办公室，要求 10 台终端联网，网络拓扑如图 2.1 所示。

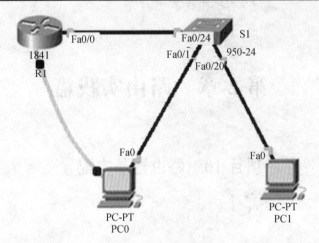

图 2.1 公司网络拓扑

5. 地址规划

该公司网络地址规划如表 2.1 所示。

表 2.1 公司网络地址规划

设备名称	设备及端口的配置地址		备注
R1	Fa0/0	192.168.1.254/24	局域网端口，连接 S1 交换机的 Fa0/24
PC0	FastEthernet	192.168.1.1/24	网关：192.168.1.254
	RS 232		连接路由器 Console 口
PC1	192.168.1.2/24		网关：192.168.1.254

注：在实际方案中，使用任意一台模块化多任务路由器完成任务，方法和配置结果一样。

6. 工作过程

步骤一：连接设备。使用网络线在工作现场按图 2.1 连接好设备。

使用 Console 线连接终端和路由器进行配置。Console 线如图 2.2 所示。

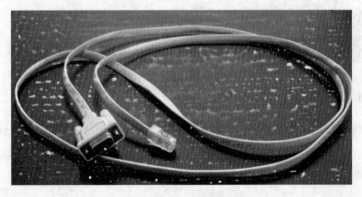

图 2.2 Console 线

第 2 章 路由实践篇

步骤二：清除配置并重启路由器。

（1）建立与路由器的终端会话。打开 PC0 的超级终端，如图 2.3 所示，建立与路由器的连接。

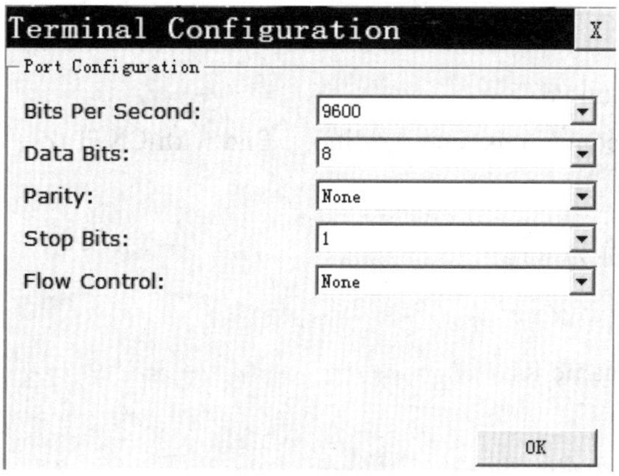

图 2.3 超级终端

（2）进入特权模式。

Router>enable
Router#

（3）清除配置。

使用 erase startup-config 命令。当收到提示 [confirm]时，按下 Enter。

Router#erase startup-config
Erasing the nvram filesystem will remove all configuration files! Continue? [confirm]
[OK]
Erase of nvram: complete
%SYS-7-NV_BLOCK_INIT: Initialized the geometry of nvram

（4）重新加载配置。

使用 reload 命令。当询问是否保存更改时，回答 NO。

Router#reload
Proceed with reload? [confirm]
System Bootstrap, Version 12.3(8r)T8, RELEASE SOFTWARE (fc1)
Cisco 1841 (revision 5.0) with 114688K/16384K bytes of memory.

Self decompressing the image :
###
[OK]
……
Continue with configuration dialog? [yes/no]: no

Press RETURN to get started!

步骤三：对路由器进行基本配置。

（1）进入全局配置模式。

Router>enable
Router#configure terminal
Enter configuration commands, one per line.   End with CNTL/Z.
Router(config)#

（2）设置路由器名称为 R1。
使用命令 Hostname。

Router(config)#hostname R1
R1(config)#

（3）禁用 DNS 查找。
使用命令 no ip domain-lookup。

R1(config)#no ip domain-lookup
R1(config)#

（4）配置执行模式口令。
使用 enable secret（或者 enable password）命令。

R1(config)#enable secret 321

（5）配置当天消息标语。
使用命令 banner motd。

R1(config)#banner motd &administrator only!&

（6）在路由器上配置控制台口令。

R1(config)#line console 0
R1(config-line)#password cisco
R1(config-line)#login

（7）为虚拟终端线路配置口令。

R1(config-line)#line vty 0 5
R1(config-line)#password class
R1(config-line)#login
R1(config-line)#exit

（8）配置 FastEthernet0/0 接口。

R1(config)#interface f0/0
R1(config-if)#ip address 192.168.1.1 255.255.255.0

R1(config-if)#no shutdown

(9) 保存 R1 配置。
使用 copy running-config startup-config 命令。

R1(config-if)#end
R1#copy running-config startup-config
Building configuration...
[OK]
R1#

步骤四：配置主机。
步骤五：验证。
(1) 使用 ping 命令，验证 PC0 到网关的连通性，PC1 到网关的连通性。
(2) 通过 PC1 虚拟终端远程登录路由器。

7. 总　　结

路由器的基本配置比较简单，主要涉及路由器名称设置、密码配置、接口配置。

项目 11　单臂路由实现 VLAN 间路由

1. 工程目标

通过路由器进行多个 VLAN 互联。

2. 技术要点

单臂路由 VLAN 间通信的原理如图 2.4 所示，路由器的以太网接口创建虚拟接口，每个虚拟接口传输一个 VLAN 的数据，同时虚拟接口是 VLAN 的网关，连接的 VLAN 是路由器的直连路由。PC1 要向 PC2 发送数据，PC1 将数据发送到自己的网关即路由上对应的虚拟子接口，路由器收到数据后查找路由表，将信息从 PC2 所在 VLAN 的虚拟子接口传出。

3. 设备清单

路由器（1 台）；网络连线（若干根）；交换机（2 台）；测试 PC（4 台）。

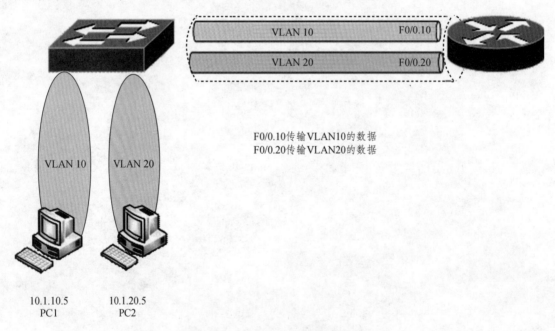

图 2.4　路由器 VLAN 路由原理

4. 工作场景

现有一新公司需要搭建网络便于管理，要求 40 台终端连入网络，现有部门行政部、销售部、财务部、后勤部,要求各部门数据相互隔离、便于管理。网络拓扑如图 2.5 所示。

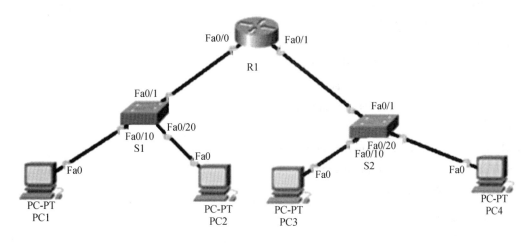

图 2.5 网络拓扑

5. 地址规划

该公司网络地址规划如表 2.2 所示。

表 2.2 网络地址规划

设备名称	设备及端口的配置地址			备注
R1	Fa0/0	Fa0/0.10	192.168.1.1/24	局域网端口,连接 S1 的 Fa0/1
		Fa0/0.20	192.168.2.1/24	
	Fa0/1	Fa0/1.30	192.168.3.1/24	局域网端口,连接 S2 的 Fa0/1
		Fa0/1.40	192.168.4.1/24	
PC1	192.168.1.2/24			网关:192.168.1.1
PC2	192.168.2.2/24			网关:192.168.2.1
PC3	192.168.3.2/24			网关:192.168.3.1
PC4	192.168.4.2/24			网关:192.168.4.1

注:在实际方案中,使用任意两台模块化多任务路由器完成任务,方法和配置结果一样。

6. 工作过程

步骤一:连接设备。使用网络线在工作现场按图 2.5 连接好设备。
步骤二:配置交换机 VLAN。
(1)配置交换机 S1 的 VLAN。

Switch>enable
Switch#configure terminal

```
Switch(config)#hostname S1
S1(config)#interface fastEthernet 0/1
S1(config-if)#switchport mode trunk
S1(config-if)#exit
S1(config)#interface fastEthernet 0/10
S1(config-if)#switchport mode access
S1(config-if)#switchport access vlan 10
S1(config-if)#exit
S1(config)#interface fastEthernet 0/20
S1(config-if)#switchport mode access
S1(config-if)#switchport access vlan 20
```

（2）配置交换机 S2 的 VLAN。

```
Switch>enable
Switch#configure terminal
Switch(config)#hostname S2
S2(config)#interface fastEthernet 0/1
S2(config-if)#switchport mode trunk
S2(config)#exit
S2(config)#interface fastEthernet 0/10
S2(config-if)#switchport mode access
S2(config-if)#switchport access vlan 30
S2(config-if)#exit
S2(config)#interface fastEthernet 0/20
S2(config-if)#switchport mode access
S2(config-if)#switchport access vlan 40
```

此时，4 台 PC 之间互相不能通信。

步骤三：配置单臂路由。

（1）配置 R1 路由器子接口地址。

```
Router>enable
Router#configure terminal
Router(config)#hostname R1
R1(config)#interface fastEthernet 0/0
R1(config-if)#no shutdown
R1(config-if)#exit
R1(config)#interface fastEthernet 0/0.10
R1(config-subif)#encapsulation dot1Q 10
R1(config-subif)#ip address 192.168.1.1 255.255.255.0
```

R1(config-subif)#exit
R1(config)#interface fastEthernet 0/0.20
R1(config-subif)#encapsulation dot1Q 20
R1(config-subif)#ip address 192.168.2.1 255.255.255.0
R1(config-subif)#exit
R1(config)#interface fastEthernet 0/1
R1(config-if)#no shutdown
R1(config-if)#exit
R1(config)#interface fastEthernet 0/1.30
R1(config-subif)#encapsulation dot1Q 30
R1(config-subif)#ip address 192.168.3.1 255.255.255.0
R1(config)#interface fastEthernet 0/1.40
R1(config-subif)#encapsulation dot1Q 40
R1(config-subif)#ip address 192.168.4.1 255.255.255.0

（2）测试连通性。此时，各 PC 之间能相互通信。

7. 总　　结

单臂路由，一个路由接口连接多个 VLAN，与交换机相连，其中要封装 Trunk 协议。要打开相应接口，并开启子接口。子接口的地址各不相同，可以看作多个不同的地址。

项目12 静态路由配置

1. 工程目标

（1）配置并激活串行接口和以太网接口。
（2）使用中间地址配置静态路由。
（3）使用送出接口配置静态路由。

2. 技术要点

路由器的主要工作就是为经过路由器的每个数据帧寻找一条最佳传输路径，并将该数据有效地传送到目的站点。由此可见，选择最佳路径的策略即路由算法是路由器的关键所在。为了完成这项工作，在路由器中保存着各种传输路径的相关数据——路径表（Routing Table），供路由选择时使用。路径表中保存着子网的标志信息、网上路由器的个数和下一个路由器的名字等内容。路径表可以是由系统管理员固定设置好的，可以由系统动态修改，可以由路由器自动调整，也可以由主机控制。

（1）静态路径表。

由系统管理员事先设置好固定的路径表称为静态（static）路径表，一般是在系统安装时就根据网络的配置情况预先设定的，它不会随未来网络结构的改变而改变。

（2）动态路径表。

动态（dynamic）路径表是路由器根据网络系统的运行情况而自动调整的路径表。路由器根据路由选择协议（Routing Protocol）提供的功能，自动学习和记忆网络运行情况，在需要时自动计算数据传输的最佳路径。

（3）配置静态路由命令。

命令格式：ip route 目的网络　目的网络掩码　下一跳地址（或者送出接口）

3. 设备清单

路由器（2台）；交换机（1台）；网络连线（若干根）；测试PC（2台）；服务器（1台）。

4. 工作场景

某小型企业在科技大厦3楼租了2间办公室，要求10台终端联网并实现公网通信。网络拓扑如图2.6所示。

第 2 章 路由实践篇

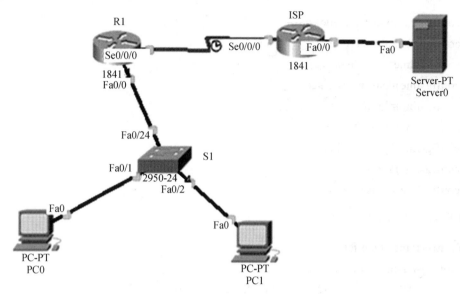

图 2.6 公司网络拓扑

5. 地址规划

该公司网络地址规划如表 2.3 所示。

表 2.3 公司网络地址规划

设备名称	设备及端口的配置地址		备注
R1	Fa0/0	192.168.1.254/24	局域网接口，连接 PC0
	S0/0/0	192.168.2.1/24	广域网接口，连接 ISP 路由器 S/0/0
ISP	Fa0/0	192.168.3.254/24	局域网接口，连接服务器
	S0/0/0	192.168.2.2/24	广域网接口，连接 R1 路由器 S/0/0
PC0	192.168.1.1/24		网关：192.168.1.254
PC1	192.168.1.2/24		网关：192.168.1.254
Server0	192.168.3.1/24		网关：192.168.3.254

注：在实际方案中，使用任意两台模块化多任务路由器完成任务，方法和配置结果一样。

6. 工作过程

步骤一：连接设备。如图 2.6 所示使用网络线在工作现场连接好设备。
步骤二：路由器基本配置。
（1）进入路由器 R1 的全局配置模式，然后配置路由器名称，禁用 DNS 查找，设置控制台和虚拟终端密码，执行密码。

Router(config)#hostname R1

R1(config)#no ip domain-lookup
R1(config)# enable secret 321
R1(config)#line console 0
R1(config-line)#password cisco
R1(config-line)#login
R1(config-line)#line vty 0 5
R1(config-line)#password class
R1(config-line)#login
R1(config-line)#exit

（2）配置路由器 R1 接口。

R1 (config)#interface fa0/0
R1 (config-if)#no shutdown
R1 (config-if)#ip address 192.168.1.254 255.255.255.0
R1 (config)#interface s0/0/0
R1 (config-if)#ip address 192.168.2.1 255.255.255.0
R1 (config-if)#clock rate 64000
R1 (config-if)#no shutdown

（3）配置路由器 ISP。

Router(config)#hostname ISP
ISP(config)#no ip domain-lookup
ISP(config)# enable secret 321
ISP(config)#line console 0
ISP(config-line)#password cisco
ISP(config-line)#login
ISP(config-line)#line vty 0 5
ISP(config-line)#password class
ISP(config-line)#login
ISP(config-line)#exit
ISP (config)#interface fa0/0
ISP (config-if)#no shutdown
ISP (config-if)#ip address 192.168.3.254 255.255.255.0
ISP (config)#interface s0/0/0
ISP (config-if)#ip address 192.168.2.2 255.255.255.0
ISP(config-if)#clock rate 64000
ISP (config-if)#no shutdown

（4）查看两台路由器的路由表。
使用命令 show ip route。此时显示的路由条目叫作直连路由。一般来讲，路由器接口直连的网络会被路由器自动添加到路由表中。

R1#show ip route
Codes: C - connected, S - static, I - IGRP, R - RIP, M - mobile, B - BGP
       D - EIGRP, EX - EIGRP external, O - OSPF, IA - OSPF inter area
       N1 - OSPF NSSA external type 1, N2 - OSPF NSSA external type 2
       E1 - OSPF external type 1, E2 - OSPF external type 2, E - EGP
       i - IS-IS, L1 - IS-IS level-1, L2 - IS-IS level-2, ia - IS-IS inter area
       * - candidate default, U - per-user static route, o - ODR
       P - periodic downloaded static route

Gateway of last resort is not set

C   192.168.1.0/24 is directly connected, FastEthernet0/0
C   192.168.2.0/24 is directly connected, Serial0/0/0

ISP#show ip route
……

C   192.168.2.0/24 is directly connected, Serial0/0/0
C   192.168.3.0/24 is directly connected, FastEthernet0/0

（5）按地址规划表配置 PC 及服务器。
此时测试 PC 和服务器连通性将会显示连接超时。
步骤三：使用下一跳地址配置静态路由。
（1）在 R1 上配置静态路由。
要使用指定的下一跳地址配置静态路由，使用以下语法：
Router(config)# ip route network-address subnet-mask ip-address
● network-address：要加入路由表的远程网络的目的网络地址。
● subnet-mask：要加入路由表的远程网络的子网掩码。可对此子网掩码进行修改，以总结一组网络。
● ip-address：一般指下一跳路由器的 IP 地址。
在 R1 路由器上，配置通往服务器网络的静态路由（使用 ISP 的 Serial 0/0/0 接口作为下一跳地址）。

R1(config)#ip route 192.168.3.0 255.255.255.0 192.168.2.2
R1(config)#

（2）查看 R1 路由表。

R1#show ip route

```
Codes: C - connected, S - static, I - IGRP, R - RIP, M - mobile, B - BGP
       D - EIGRP, EX - EIGRP external, O - OSPF, IA - OSPF inter area
       N1 - OSPF NSSA external type 1, N2 - OSPF NSSA external type 2
       E1 - OSPF external type 1, E2 - OSPF external type 2, E - EGP
       i - IS-IS, L1 - IS-IS level-1, L2 - IS-IS level-2, ia - IS-IS inter area
       * - candidate default, U - per-user static route, o - ODR
       P - periodic downloaded static route

Gateway of last resort is not set

C    192.168.1.0/24 is directly connected, FastEthernet0/0
C    192.168.2.0/24 is directly connected, Serial0/0/0
S    192.168.3.0/24 [1/0] via 192.168.2.2
```

（3）在 ISP 上配置静态路由。

```
ISP(config)#ip route 192.168.1.0 255.255.255.0 192.168.2.1
ISP(config)#end
ISP#show ip route
Codes: C - connected, S - static, I - IGRP, R - RIP, M - mobile, B - BGP
       D - EIGRP, EX - EIGRP external, O - OSPF, IA - OSPF inter area
       N1 - OSPF NSSA external type 1, N2 - OSPF NSSA external type 2
       E1 - OSPF external type 1, E2 - OSPF external type 2, E - EGP
       i - IS-IS, L1 - IS-IS level-1, L2 - IS-IS level-2, ia - IS-IS inter area
       * - candidate default, U - per-user static route, o - ODR
       P - periodic downloaded static route

Gateway of last resort is not set

S    192.168.1.0/24 [1/0] via 192.168.2.1
C    192.168.2.0/24 is directly connected, Serial0/0/0
C    192.168.3.0/24 is directly connected, FastEthernet0/0
```

（4）使用 ping 检查主机 PC0 与服务器之间的连通性。此处结果应该连接成功。

步骤四：使用送出接口配置静态路由。

（1）在 R1 上配置静态路由。

要使用指定的送出接口配置静态路由，使用以下语法：

Router(config)# ip route network-address　subnet-mask　exit-interface

- network-address—要加入路由表的远程网络的目的网络地址。
- subnet-mask—要加入路由表的远程网络的子网掩码。可对此子网掩码进行修改，以总结一组网络。

- exit-interface—将数据包转发到目的网络时使用的传出接口。

R1(config)# ip route 192.168.3.0 255.255.255.0 Serial0/0/0
R1(config)#end
R1#show ip route
Codes: C - connected, S - static, I - IGRP, R - RIP, M - mobile, B - BGP
       D - EIGRP, EX - EIGRP external, O - OSPF, IA - OSPF inter area
       N1 - OSPF NSSA external type 1, N2 - OSPF NSSA external type 2
       E1 - OSPF external type 1, E2 - OSPF external type 2, E - EGP
       i - IS-IS, L1 - IS-IS level-1, L2 - IS-IS level-2, ia - IS-IS inter area
       * - candidate default, U - per-user static route, o - ODR
       P - periodic downloaded static route

Gateway of last resort is not set

C   192.168.1.0/24 is directly connected, FastEthernet0/0
C   192.168.2.0/24 is directly connected, Serial0/0/0
S   192.168.3.0/24 is directly connected, Serial0/0/0

注意：请比较此处路由表条目和使用下一跳地址配置静态路由时 R1 的路由条目。

（2）在 ISP 上配置静态路由。

ISP(config)# ip route 192.168.1.0 255.255.255.0 Serial0/0/0
ISP(config)#

（3）验证 PC 和服务器的连通性。

7. 总　结

当网络中出现非直接相连目的地网络时，可以对路由器添加静态路由来指向相应的非直接相连目的地。直连路由会自动被路由器添加到路由表中。

添加静态路由有两种方法：给出下一跳地址或者给出送出接口。下一跳地址是直连路由的接口地址，送出接口是路由器自己的接口。

项目 13 默认路由配置

1. 工程目标

（1）配置默认静态路由。
（2）配置总结静态路由。

2. 技术要点

默认路由和静态路由的命令格式一样，只是把目的地 IP 和子网掩码改成 0.0.0.0 和 0.0.0.0。总结路由要找到需要总结的几个网络的相同部分，子网掩码的计算方法是相同部分置 1，不同部分置 0。

3. 设备清单

路由器（2 台）；网络连线（若干根）；测试 PC（2 台）；交换机（2 台）。

4. 工作场景

某小型企业在科技大厦 3 楼租了 2 间办公室，要求 10 台终端联网并实现公网通信。网络拓扑如图 2.7 所示。

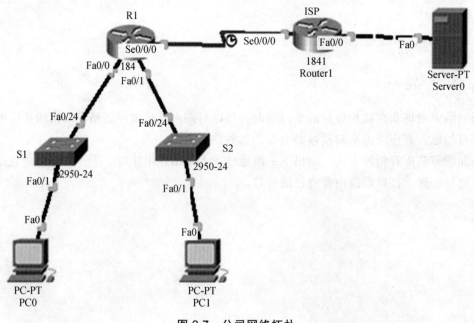

图 2.7 公司网络拓扑

5. 地址规划

该公司网络地址规划如表 2.4 所示。

表 2.4 公司网络地址规划

设备名称	设备及端口的配置地址		备注
R1	Fa0/0	192.168.1.254/24	局域网接口，连接 S1 的 f0/24
	Fa0/1	192.168.2.254/24	局域网接口，连接 S2 的 f0/24
	S0/0/0	192.168.4.1/24	广域网接口，连接 ISP 路由器 S/0/0
ISP	Fa0/0	192.168.3.254/24	局域网接口，连接服务器
	S0/0/0	192.168.4.2/24	广域网接口，连接 R1 路由器 S/0/0
PC0	192.168.1.1/24		网关：192.168.1.254
PC1	192.168.2.1/24		网关：192.168.2.254
Server0	192.168.3.1/24		网关：192.168.3.254

注：在实际方案中，使用任意两台模块化多任务路由器完成任务，方法和配置结果一样。

6. 工作过程

步骤一：连接设备。使用网络线在工作现场按图 2.7 连接好设备。

步骤二：路由器基本配置。

（1）进入路由器 R1 的全局配置模式，然后配置路由器名称，禁用 DNS 查找，设置控制台和虚拟终端密码，执行密码。

Router(config)#hostname R1
R1(config)#no ip domain-lookup
R1(config)#enable secret 321
R1(config)#line console 0
R1(config-line)#password cisco
R1(config-line)#login
R1(config-line)#line vty 0 5
R1(config-line)#password class
R1(config-line)#login
R1(config-line)#exit

（2）配置路由器 R1 接口。

R1 (config)#interface fa0/0
R1 (config-if)#no shutdown
R1 (config-if)#ip address 192.168.1.254 255.255.255.0
R1 (config)#interface fa0/1

R1 (config-if)#no shutdown
R1 (config-if)#ip address 192.168.2.254 255.255.255.0
R1 (config)#interface s0/0/0
R1 (config-if)#ip address 192.168.4.1 255.255.255.0
R1(config-if)#clock rate 64000
R1 (config-if)#no shutdown

（3）配置路由器 ISP。

Router(config)#hostname ISP
ISP(config)#no ip domain-lookup
ISP(config)#enable secret 321
ISP(config)#line console 0
ISP(config-line)#password cisco
ISP(config-line)#login
ISP(config-line)#line vty 0 5
ISP(config-line)#password class
ISP(config-line)#login
ISP(config-line)#exit
ISP (config)#interface fa0/0
ISP (config-if)#no shutdown
ISP (config-if)#ip address 192.168.3.254 255.255.255.0
ISP (config)#interface s0/0/0
ISP (config-if)#ip address 192.168.4.2 255.255.255.0
ISP(config-if)#clock rate 64000
ISP (config-if)#no shutdown

步骤三：使用默认静态路由。

在上节中，我们通过配置下一跳和送出接口已经为路由器配置了通往特定目的地的具体路由。但是 Internet 上的网络我们能全部添加到 R1 的路由表里吗？显然这样做工作量会相当大，根本不可能实现。为了缩小路由表的大小，这里我们使用默认静态路由。当路由器没有更好、更精确的路由可到达目的地时，它就会使用默认静态路由。

要配置默认静态路由，使用以下命令：

Router(config)#ip route 0.0.0.0 0.0.0.0　　{ ip-address| interface}

（1）为 R1 配置默认路由。

R1(config)#ip route 0.0.0.0 0.0.0.0 192.168.4.2

（2）查看路由表。

R1#show ip route
Codes: C - connected, S - static, I - IGRP, R - RIP, M - mobile, B - BGP

D - EIGRP, EX - EIGRP external, O - OSPF, IA - OSPF inter area
N1 - OSPF NSSA external type 1, N2 - OSPF NSSA external type 2
E1 - OSPF external type 1, E2 - OSPF external type 2, E - EGP
i - IS-IS, L1 - IS-IS level-1, L2 - IS-IS level-2, ia - IS-IS inter area
* - candidate default, U - per-user static route, o - ODR
P - periodic downloaded static route

Gateway of last resort is 192.168.4.2 to network 0.0.0.0

C   192.168.1.0/24 is directly connected, FastEthernet0/0
C   192.168.2.0/24 is directly connected, Serial0/0/0
S*   0.0.0.0/0 [1/0] via 192.168.4.2

步骤四：配置总结静态路由。

我们可以在 ISP 上配置两条静态路由来指向 192.168.1.0/24 和 192.168.2.0/24 网段。

由于这些网络彼此非常接近，所以可将它们总结为一条路由。此方法同样可缩小路由表的大小，从而使得路由查找过程更有效率。

观察以二进制形式表示的这两个网络，它们的靠左 22 位完全相同。

192.168.1.0 11000000.10101000.00000001.00000000
192.168.2.0 11000000.10101000.00000010.00000000

如果不考虑这 22 位之后的其余位，我们可以将这两个网络总结为 192.168.0.0。

为了屏蔽靠左前 22 位，我们使用靠左 22 位为全 1 的掩码：

11111111.11111111.11111100.00000000 即 255.255.252.0

（1）为 ISP 配置总结静态路由。

ISP (config)#ip route 192.168.0.0 255.255.252.0 192.168.4.1

（2）查看路由表。

ISP #show ip route
Codes: C - connected, S - static, I - IGRP, R - RIP, M - mobile, B - BGP
       D - EIGRP, EX - EIGRP external, O - OSPF, IA - OSPF inter area
       N1 - OSPF NSSA external type 1, N2 - OSPF NSSA external type 2
       E1 - OSPF external type 1, E2 - OSPF external type 2, E - EGP
       i - IS-IS, L1 - IS-IS level-1, L2 - IS-IS level-2, ia - IS-IS inter area
       * - candidate default, U - per-user static route, o - ODR
       P - periodic downloaded static route

Gateway of last resort is not set

S   192.168.0.0/22 [1/0] via 192.168.4.1
C   192.168.3.0/24 is directly connected, FastEthernet0/0

C   192.168.4.0/24 is directly connected, Serial0/0/0

步骤五：配置主机及服务器地址。
步骤六：测试连通性。此时，PC 与服务器之间能相互通信。

7. 总　结

默认路由是对 IP 数据包中的目的地址找不到存在的其他路由时，路由器所选择的路由。合理配置默认路由可以简化工作量，提高路由器转发速度。

总结路由是当目的网络地址相近的网络对应下一跳是同一出口时，可以将相近的几个网络地址进行总结，得出一条包含这几个网络的总结地址。

项目 14 RIP 路由协议应用

1. 工程目标

（1）理解 RIP 协议的基本特性。
（2）掌握 RIP 协议的配置。

2. 技术要点

1）协议概述

路由信息协议（Routing Information Protocol，RIP）是一种使用最广泛的内部网关协议（IGP）。IGP 是在内部网络上使用的路由协议(在少数情形下,也可以用于连接到因特网的网络)，它可以通过不断的交换信息让路由器动态的适应网络连接的变化，这些信息包括每个路由器可以到达哪些网络以及这些网络有多远等。IGP 是应用层协议，并使用 UDP 作为传输协议。

2）协议特点

- RIP 是自治系统内部使用的协议，即内部网关协议，使用的是距离矢量算法。
- RIP 使用 UDP 的 520 端口进行 RIP 进程之间的通信。
- RIP 主要有两个版本：RIPv1 和 RIPv2。RIPv1 协议的具体描述在 RFC1058 中，RIPv2 是对 RIPv1 协议的改进。
- RIP 协议以跳数作为网络度量值。
- RIP 协议采用广播或组播进行路由更新，其中 RIPv1 使用广播，而 RIPv2 使用组播（224.0.0.9）。
- RIP 协议支持主机被动模式，即 RIP 协议允许主机只接收和更新路由信息而不发送信息。
- RIP 协议支持默认路由传播。
- RIP 协议的网络直径不超过 15 跳，适合于中小型网络。16 跳时认为网络不可达，不适合大型网络。
- RIPv1 是有类路由协议，RIPv2 是无类路由协议，即 RIPv2 的报文中含有掩码信息。

3）工作原理

同一自治系统（A.S.）中的路由器每 30 s 会与相邻的路由器交换子讯息，以动态地建立路由表。
RIP 允许最大的 hop 数（跳数）为 15，多于 15 跳认为网络不可达。

4）协议版本

RIP 共有三个版本，RIPv1、RIPv2、RIPng。

其中 RIPV1 和 RIPV2 是用在 IPV4 的网络环境里，RIPng 是用在 IPV6 的网络环境里。

- RIPv1

RIPv1 使用分类路由，定义在[RFC 1058]中。在它的路由更新（Routing Updates）中并不带有子网的资讯，因此它无法支持可变长度的子网掩码。这个限制造成在 RIPv1 的网络中，同级网络无法使用不同的子网掩码。换句话说，在同一个网络中所有的子网络数目都是相同的。另外，它也不支持对路由过程的认证，使得 RIPv1 有一些轻微的缺陷，有被攻击的可能。

- RIPv2

因为 RIPv1 的缺陷，RIPv2 在 1994 年被提出，将子网络的资讯包含在内，通过这样的方式提供无类别域间路由，不过最大节点数 15 的这个限制仍然被保留着。另外针对安全性的问题，RIPv2 也提供一套方法，通过加密来达到认证的效果。而之后[RFC 2082][3]也定义了利用 MD5 来达到认证的方法。 RIPv2 的相关规定在[RFC 2453][4]] orSTD56 中。

现今的 IPv4 网络中使用的大多是 RIPv2，RIPv2 是在 RIPv1 的基础上改进的路由信息协议，RIPv2 和 RIPv1 相比主要有如表 2.5 所示的区别。

表 2.5　RIPv1 与 RIPv2 比较

RIPv1	RIPv2
数据包中不包含子网掩码	数据包中包含子网掩码
采用广播更新路由条目	采用组播 224.0.0.9 更新路由条目
不提供验证	支持明文和 MD5 验证，更高安全性
有类路由协议	无类路由协议

- RIPng

主要是针对 IPv6 做一些延伸的规范。

5）路由环路解决方法

- 记数最大值（maximum hop count）：定义最大跳数（最大为 15 跳），当跳数为 16 时，目标为不可达。
- 水平分割（split horizon）：从一个接口学习到的路由不会再广播回该接口。Cisco 可以对每个接口关闭水平分割功能。
- 路由毒化（route posion）：当拓扑变化时，路由器会将失效的路由标记为 possibly down 状态，并分配一个不可达的度量值。
- 毒性逆转（poison reverse）：从一个接口学习的路由会发送回该接口，但是已经被毒化，跳数设置为 16 跳，不可达。
- 触发更新（trigger update）：一旦检测到路由崩溃，立即广播路由刷新报文，而不会等到下一刷新周期。
- 抑制计时器（holddown timer）：防止路由表频繁翻动，增加了网络的稳定性。

3. 设备清单

路由器（3 台）；网络连线（若干根）；测试 PC（2 台）。

4. 工作场景

如图 2.8 所示模拟的是四川信息职业技术学院雪峰和东坝两个分散校区场景，使用 3 台路由器表示分散校园网络中 2 个不同园区区域：左边路由器连接的是雪峰校区，右边路由器连接的是东坝校区。现使用网络线路直接连接，希望通过 RIPv2 动态路由技术，实现分散校园网络系统互连互通。

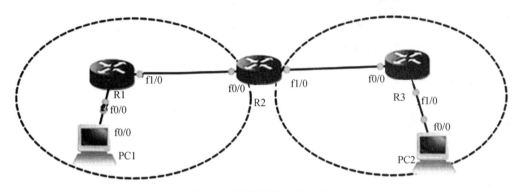

图 2.8 区域网络工作场景

5. 地址规划

两个分散校园园区网络地址规划如表 2.6 所示。

表 2.6 两个分散校园园区网络地址规划

设备名称	设备及端口的配置地址		备注
R1	Fa0/0	172.16.1.1/24	局域网端口，连接 PC1
	Fa1/0	172.16.2.1/24	局域网端口，连接 R2 路由器 Fa0/0
R2	Fa0/0	172.16.2.2/24	局域网端口，连接 R1 路由器 Fa1/0
	Fa1/0	172.17.1.1/24	局域网端口，连接 R3 路由器 Fa0/0
R3	Fa0/0	172.17.1.2/24	局域网端口，连接 R2 路由器 Fa1/0
	Fa1/0	172.17.2.1/24	局域网端口，连接 PC2
PC1	172.16.1.2/24		网关：172.16.1.1
PC2	172.17.2.2/24		网关：172.17.2.1

注：在实际方案中，使用任意两台模块化多任务路由器完成任务，方法和配置结果一样。

6. 工作过程

步骤一：连接设备。按图 2.8 使用网络线缆在工作现场连接好设备。
步骤二：配置路由器设备基本接口信息。
（1）配置 R1 路由器端口的地址。

Router(config)#hostname R1

R1#configure terminal
R1 (config)#interface fa0/0
R1 (config-if)#no shutdown
R1 (config-if)#ip address 172.16.1.1 255.255.255.0
R1 (config)#interface fa1/0
R1 (config-if)#ip address 172.16.2.1 255.255.255.0
R1 (config-if)#no shutdown

（2）配置 R2 路由器的端口地址。

Router(config)#hostname R2
R2#configure terminal
R2 (config)#interface fa0/0
R2 (config-if)#ip address 172.16.2.2 255.255.255.0
R2 (config-if)#no shutdown
R2 (config)#interface fa1/0
R2 (config-if)#ip address 172.17.1.1 255.255.255.0
R2 (config-if)#no shutdown

（3）配置 R3 路由器的端口地址。

Router(config)#hostname R3
R3#configure terminal
R3 (config)#interface fa 0/0
R3 (config-if)#ip address 172.17.1.2 255.255.255.0
R3 (config-if)#no shutdown
R3 (config)#interface fa 1/0
R3 (config-if)#ip address 172.17.2.1 255.255.255.0
R3 (config-if)#no shutdown

步骤三：配置测试计算机地址，测试网络连通性。

（1）配置 PC1 的地址为 172.16.1.2/24，网关为 172.16.1.1；配置 PC2 的地址为 172.17.2.2/24，网关为 172.17.2.1。

（2）使用 ping 命令测试网络连通性，PC1 无法和 PC2 进行通信。

（3）查询网络不能通信的原因，查看 R1 路由表，缺少到 172.17.1.0/24 和 172.17.2.0/24 网络的路由。

R1#show ip route
Codes: C - connected, S - static, R - RIP, M - mobile, B - BGP
    D - EIGRP, EX - EIGRP external, O - OSPF, IA - OSPF inter area
    N1 - OSPF NSSA external type 1, N2 - OSPF NSSA external type 2
    E1 - OSPF external type 1, E2 - OSPF external type 2

i - IS-IS, su - IS-IS summary, L1 - IS-IS level-1, L2 - IS-IS level-2
ia - IS-IS inter area, * - candidate default, U - per-user static route
o - ODR, P - periodic downloaded static route
Gateway of last resort is not set

    172.17.0.0/24 is subnetted, 1 subnets
C   172.17.1.0 is directly connected, FastEthernet0/0
    172.16.0.0/24 is subnetted, 1 subnets
C   172.16.2.0 is directly connected, FastEthernet1/0

步骤四：配置网络的动态 RIPv2 路由，实现网络的连通。

（1）配置 R1 路由器到达 172.17.1.0/24 和 172.17.2.0/24 网络的动态 RIP 路由。

R1#configure terminal
R1(config)#router rip
R1(config-router)#version 2   //启用第 2 版本
R1(config-router)#no auto-summary   //关闭自动汇总
R1(config-router)#network 172.16.1.0   //宣告直连网络地址
R1(config-router)#network 172.16.2.0

（2）配置 R2 路由器到达 172.16.1.0/24 和 172.17.2.0/24 网络的动态 RIP 路由。

R2(config)#router rip
R2(config-router)#version 2
R2(config-router)#no auto-summary
R2(config-router)#network 172.16.1.0
R2(config-router)#network 172.17.1.0

（3）配置 R3 路由器到达 172.16.1.0/24 和 172.12.2.0/24 网络的动态 RIP 路由。

R3(config)#router rip
R3(config-router)#version 2
R3(config-router)#no auto-summary
R3(config-router)#network 172.17.1.0
R3(config-router)#network 172.17.2.0

（4）查看 R1 的路由表，通过 RIP 路由技术学习到达 172.17.1.0/24 和 172.17.1.0/24 网络的路由。

R1#show ip route
Codes: C - connected, S - static, R - RIP, M - mobile, B - BGP
      D - EIGRP, EX - EIGRP external, O - OSPF, IA - OSPF inter area
      N1 - OSPF NSSA external type 1, N2 - OSPF NSSA external type 2
      E1 - OSPF external type 1, E2 - OSPF external type 2

```
              i - IS-IS, su - IS-IS summary, L1 - IS-IS level-1, L2 - IS-IS level-2
       ia - IS-IS inter area, * - candidate default, U - per-user static route
              o - ODR, P - periodic downloaded static route
Gateway of last resort is not set

        172.17.0.0/24 is subnetted, 2 subnets
R       172.17.1.0 [120/1] via 172.16.2.2, 00:00:00, FastEthernet1/0
R       172.17.2.0 [120/2] via 172.16.2.2, 00:00:00, FastEthernet1/0
        172.16.0.0/24 is subnetted, 2 subnets
C       172.16.1.0 is directly connected, FastEthernet0/0
C       172.16.2.0 is directly connected, FastEthernet1/0
```

注：120为距离管理值，1和2为度量值（跳数），RIP以跳数为度量值参数用于衡量到达目的地的距离。

7. 总　结

RIP是距离矢量路由选择协议，该协议简单易行，但该协议过于简单，以跳数作为基数依据确定度量值，选出的路径并非最佳路径；最大跳数是15跳，如果大于15跳，它就会丢弃数据包，不适合大型网络结构；收敛速度慢，可靠性非常差；定期更新路由信息占用网络带宽很大。

项目 15　配置 OSPF 路由实现区域网连通

1. 工程目标

（1）理解 OSPF 协议工作原理。
（2）掌握通过动态路由 OSPF 方式实现区域网的连通。

2. 技术要点

1）协议概述

开放式最短路径优先（Open Shortest Path First，OSPF）是一个内部网关协议（Interior Gateway Protocol，IGP），是一种典型的链路状态（Link-state）的路由协议，用于在单一自治系统（Autonomous System，AS）内决策路由。OSPF 通过路由器之间通告网络接口的状态来建立链路状态数据库，生成最短路径树，每个 OSPF 路由器使用这些最短路径构造路由表。

2）OSPF 的数据包

（1）数据包类型。
- HELLO 12 Database Description 数据库的描述 DBD 可靠；
- Link-state Request 链路状态请求包 LSR 可靠；
- Link-state Update 链路状态更新包 LSU 可靠；
- Link-state Acknowlegment 链路状态确认包 LSACK。

（2）OSPF 的 Hello 协议作用。
- 用于发现邻居；
- 在成为邻居之前,必须对 Hello 包里的一些参数协商成功；
- Hello 包在邻居之间扮演着 keepalive 的角色；
- 允许邻居之间的双向通信；
- 它在 NBMA（Nonbroadcast Multi-access）网络上选举 DR 和 BDR（NBMA 中默认 30 s 发送一次，多路访问和点对点网络上默认 10 s 发送一次）。

（3）Hello Packet 包含信息。
- 源路由器的 RID；
- 源路由器的 Area ID；
- 源路由器接口的掩码；
- 源路由器接口的认证类型和认证信息；
- 源路由器接口的 Hello 包发送的时间间隔；
- 源路由器接口的无效时间间隔；
- 优先级；

- DR/BDR；
- 五个标记位（flag bit）；
- 源路由器的所有邻居的 RID。

3）OSPF 路由器类型

（1）内部路由器：OSPF 路由器上所有直连的链路都处于同一个区域。
（2）主干路由器：具有连接区域 0 接口的路由器。
（3）区域边界路由器（ABR）：路由器与多个区域相连的路由器。
（4）自制系统边界路由器（ASBR）：与 AS 外部的路由器相连并互相交换路由信息的路由器。

4）OSPF 区域

区域长度 32 位，可以用 10 进制，也可以类似于 IP 地址的点分十进制分三种通信量。

（1）Intra-Area Traffic：域内间通信量；
（2）Inter-Area Traffic：域间通信量；
（3）External Traffic：外部通信量。

5）虚链路（Virtual Link）

以下 2 种情况需要使用到虚链路：
（1）通过一个非骨干区域连接到一个骨干区域。
（2）通过一个非骨干区域连接一个分段的骨干区域两边的部分区域。
虚链接是一个逻辑的隧道（Tunnel），配置虚链接的一些规则：
- 虚链接必须配置在 2 个 ABR 之间。
- 虚链接所经过的区域叫 Transit Area，它必须拥有完整的路由信息。
- Transit Area 不能是 Stub Area。
- 尽量避免使用虚链接，它增加了网络的复杂程度和加大了排错的难度。

3. 设备清单

路由器（2 台）；网络连线（若干根）；测试 PC（2 台）。

4. 工作场景

如图 2.9 所示模拟的是某中学分散两个校区场景，使用 2 台路由器表示分散校园网络中 2 个不同园区区域：左边路由器连接是新校区，右边路由器连接是老校区。现使用网络线路直接连接，希望通过 OSPF 动态路由技术，实现分散校园网络系统互连互通。

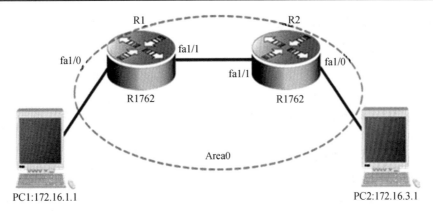

图 2.9 区域网络工作场景

5. 地址规划

两个分散校园园区网络地址规划如表 2.7 所示。

表 2.7 两个分散校园园区网络地址规划

设备名称	设备及端口的配置地址		备注
R1	Fa1/0	172.16.1.1/24	局域网端口，连接 PC1
	Fa1/1	172.16.2.1/24	局域网端口，连接 R2 路由器 Fa1/0
R2	Fa1/0	172.16.2.2/24	局域网端口，连接 R1 路由器 Fa1/1
	Fa1/1	172.16.3.1/24	局域网端口，连接 PC2
PC1	172.16.1.2/24		网关：172.16.1.1
PC2	172.16.3.2/24		网关：172.16.3.1

注：在实际方案中，使用任意两台模块化多任务路由器完成任务，方法和配置结果一样。

6. 工作过程

步骤一：连接设备。按图 2.9 使用网络线缆在工作现场连接好设备。
步骤二：配置路由器设备基本接口信息。
（1）配置 R1 路由器端口的地址。

Router(config)#hostname R1
R1#configure terminal
R1 (config)#interface fa1/0
R1 (config-if)#no shutdown
R1 (config-if)#ip address 172.16.1.1 255.255.255.0
R1 (config)#interface fa1/1
R1 (config-if)#ip address 172.16.2.1 255.255.255.0
R1 (config-if)#no shutdown

（2）配置 R2 路由器的端口地址。

Router(config)#hostname R2
R2#configure terminal
R2 (config)#interface fa1/1
R2(config-if)#ip address 172.16.2.2 255.255.255.0
R2 (config-if)#no shutdown
R2 (config)#interface fa1/0
R2 (config-if)#ip address 172.16.3.1 255.255.255.0
R2 (config-if)#no shutdown

步骤三：配置测试计算机地址，测试网络连通性。

（1）配置 PC1 的地址为 172.16.1.2/24，网关为 172.16.1.1；配置 PC2 的地址为 172.16.3.2/24，网关为 172.16.3.1。

（2）使用 ping 命令测试网络连通性，PC1 无法和 PC2 进行通信。

（3）查询网络不能通信原因，查看 R1 路由表，缺少到 172.16.3.0/24 网络的路由。

R1#show ip route
Code:C-connected, S-static, R-RIP
　　　　O-OSPF, IA-OSPF inter area
　　　　N1-OSPF NSSA external type 1, N2-OSPF NSSA external type 2
　　　　E1-OSPF external type 1, E2-OSPF external type 2
　　　　*-candidate default
Gateway of last resort is no set
C　　　172.16.2.0/24 is directly connected, FastEthernet1/1
C　　　172.16.2.1/24 is local host.

步骤四：配置网络的动态 OSPF 路由，实现网络的连通。

（1）配置 R1 路由器到达 172.16.3.0/24 网络的动态 OSPF 路由。

R1#configure terminal
R1 (config)#router ospf
R1 (config)#router-id 1.1.1.1
R1 (config-router)#network 172.16.1.0 0.0.0.255 area 0
R1 (config-router)#network 172.16.2.0 0.0.0.255 area 0
R1 (config-router)#exit
R1 (config)#

（2）配置 R2 路由器到达 172.16.1.0/24 网络的动态 OSPF 路由。

R2#configure terminal
R2 (config)#router ospf
R2 (config)#router-id 2.2.2.2

R2 (config-router)#network 172.16.2.0 0.0.0.255 area 0

R2 (config-router)#network 172.16.3.0 0.0.0.255 area 0

R2 (config-router)#exit

R2 (config)#

注意：本项目做的是单区域的配置，区域编号为 0，为主区域。

（3）查看 R1 的路由表，通过 OSPF 路由技术学习到达 172.16.3.0/24 网络的路由。

R1#show ip route

Code:C-connected, S-static, R-RIP

  O-OSPF, IA-OSPF inter area

  N1-OSPF NSSA external type 1, N2-OSPF NSSA external type 2

  E1-OSPF external type 1, E2-OSPF external type 2

  *-candidate default

Gateway of last resort is no set

C 172.16.2.0/24 is directly connected, FastEthernet1/1

C 172.16.2.1/24 is local host.

O 172.16.3.0/24 [110/1] via 172.16.2.2, 00:00:16, FastEthernet 1/1

7. 总　　结

OSPF 作为一种内部网关协议，用于在同一个自治系统（AS）中的路由器之间交换路由信息。它具有以下特征：

- 适用大规模网络；
- 收敛速度快；
- 无网络环路；
- 支持 VLSM 和 CIDR；
- 支持等价路由；
- 采用 COST 作为度量标准；
- 维护邻居表、拓扑表和路由表。

项目 16　路由重分布

1. 工程目标

（1）掌握协议度量值的配置。
（2）掌握路由重分布参数配置。
（3）掌握静态路由重分布配置。
（4）学会 RIP 和 OSPF 重分布配置。

2. 技术要点

路由重分布为同一个互联网络中支持多种路由协议，执行路由重分布的路由器被称为边界路由器。路由器重分布时计量单位和管理距离是必要参数，每一种路由协议都有自己的度量标准，为了协议兼容性，所以在进行路由重分布时必须转换度量标准。度量值的定义在路由重分布里，它是一条从外部重分布进来的路由的初始度量值，其中 RIP 的度量值为无限大，OSPF 的度量值为 20。

3. 设备清单

路由器（4 台）；网络连线（若干根）；测试 PC（3 台）。

4. 工作场景

如图 2.10 所示模拟的是某大学分散两个校区场景，使用 2 台路由器表示分散校园网络中 2 个不同园区区域：左边路由器连接是东校区，使用 OSPF 通信；右边路由器连接是西校区，使用 RIP 通信。除 R0 与 ISP 互联外，R0 到外网使用默认路由出去，实现分散校园网络系统和外网服务器互连互通。

5. 地址规划

两个分散校园园区网络地址规划如表 2.8 所示。

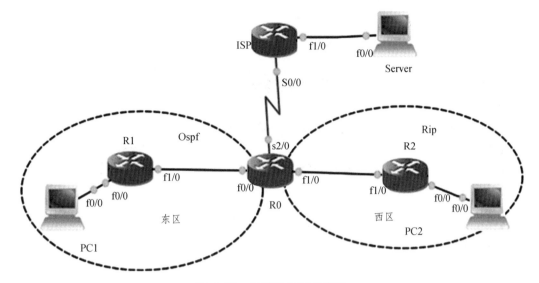

图 2.10 区域网络工作场景

表 2.8 两个分散校园园区网络地址规划

设备名称	设备及端口的配置地址		备 注
R0	Fa0/0	172.16.2.2/24	局域网端口，连接 R1 路由器 Fa1/0
	Fa1/0	172.17.2.2/24	局域网端口，连接 R2 路由器 Fa1/0
	S2/0	100.1.1.1/24	广域网端口，连接 ISP 路由器 S0/0
R1	Fa0/0	172.16.1.1/24	局域网端口，连接 PC1
	Fa1/0	172.16.2.1/24	局域网端口，连接 R0 路由器 Fa0/0
R2	Fa0/0	172.17.1.1/24	局域网端口，连接 PC2
	Fa1/1	172.17.2.1/24	局域网端口，连接 R0 路由器 Fa1/0
ISP	S0/0	100.1.1.2/24	广域网端口，连接 R0 路由器 S0/0
	Fa1/0	200.1.1.1/24	局域网端口，连接 Server
PC1	172.16.1.2/24		网关：172.16.1.1
PC2	172.17.1.2/24		网关：172.17.1.1
Server	200.1.1.2/24		网关：200.1.1.1

注：在实际方案中，使用任意两台模块化多任务路由器完成任务，方法和配置结果一样。

6. 工作过程

步骤一：连接设备。按图 2.10 使用网络线缆在工作现场连接好设备。

步骤二：配置路由器设备基本接口信息。

（1）配置 R1 路由器端口的地址。

R1>enable

R1(config)#interface fastEthernet 0/0

R1(config-if)#no shutdown

R1(config-if)#ip address 172.16.1.1 255.255.255.0

R1(config-if)#exit

R1(config)#interface fastEthernet 1/0

R1(config-if)#no shutdown

R1(config-if)#ip address 172.16.2.1 255.255.255.0

R1(config-if)#end

（2）配置 R0 路由器端口的地址。

R0>enable

R0(config)#interface fa

R0(config)#interface fastEthernet 0/0

R0(config-if)#no shutdown

R0(config-if)#ip address 172.16.2.2 255.255.255.0

R0(config-if)#exit

R0(config)#interface fastEthernet 1/0

R0(config-if)#no shutdown

R0(config-if)#ip address 172.17.2.2 255.255.255.0

R0(config-if)#exit

R0(config)#interface serial 2/0

R0(config-if)#no shutdown

R0(config-if)#ip address 100.1.1.1 255.255.255.0

R0(config-if)#end

（3）配置 R2 路由器端口的地址。

R2>enable

R2#configure terminal

R2(config)#interface fastEthernet 0/0

R2(config-if)#no shutdown

R2(config-if)#ip address 172.17.1.1 255.255.255.0

R2(config-if)#exit

R2(config)#interface fastEthernet 1/0

R2(config-if)#no shutdown

R2(config-if)#ip address 172.17.2.1 255.255.255.0

R2(config-if)#end

（4）配置 ISP 路由器端口的地址。

ISP>enable

ISP #configure terminal

ISP (config)#interface serial 0/0

ISP (config-if)#no shutdown

ISP (config-if)#ip address 100.1.1.2 255.255.255.0

ISP (config-if)#exit

ISP (config)#interface fastEthernet 1/0

ISP (config-if)#no shutdown

ISP (config-if)#ip address 200.1.1.1 255.255.255.0

ISP (config-if)#exit

步骤三：配置 OSPF 协议让东区互通。

（1）在 R1 上启用 OSPF 协议。

R1(config)#router ospf 1

R1(config-router)#router-id 1.1.1.1

R1(config-router)#network 172.16.1.0 0.0.0.255 area 0

R1(config-router)#network 172.16.2.0 0.0.0.255 a

R1(config-router)#network 172.16.2.0 0.0.0.255 area 0

R1(config-router)#

（2）在 R1 上启用 OSPF 协议。

R1(config)#router ospf 1

R1(config-router)#router-id 2.2.2.2

R1(config-router)#network 172.16.2.0 0.0.0.255 area 0

（3）查看 R0 上路由表。

R0#show ip route

Codes: C - connected, S - static, R - RIP, M - mobile, B - BGP

       D - EIGRP, EX - EIGRP external, O - OSPF, IA - OSPF inter area

       N1 - OSPF NSSA external type 1, N2 - OSPF NSSA external type 2

       E1 - OSPF external type 1, E2 - OSPF external type 2

       i - IS-IS, su - IS-IS summary, L1 - IS-IS level-1, L2 - IS-IS level-2

ia - IS-IS inter area, * - candidate default, U - per-user static route

       o - ODR, P - periodic downloaded static route

Gateway of last resort is not set

     100.0.0.0/24 is subnetted, 1 subnets

C   100.1.1.0 is directly connected, Serial2/0

     172.17.0.0/24 is subnetted, 1 subnets

C   172.17.2.0 is directly connected, FastEthernet1/0

172.16.0.0/24 is subnetted, 2 subnets
O   172.16.1.0 [110/2] via 172.16.2.1, 00:00:46, FastEthernet0/0
C   172.16.2.0 is directly connected, FastEthernet0/0

步骤四：配置 RIP 协议让西区互通。

（1）在 R2 上启用 RIP 协议。

R2(config)#router rip
R2(config-router)#no auto-summary
R2(config-router)#version 2
R2(config-router)#network 172.17.1.0
R2(config-router)#network 172.17.2.0
R2(config-router)#

（2）在 R0 上启用 RIP 协议。

R0(config)#router rip
R0(config-router)#version 2
R0(config-router)#no auto-summary
R0(config-router)#net
R0(config-router)#network 172.17.2.0

（3）查看 R0、R1 和 R2 上路由表。

R0#show ip route
Codes: C - connected, S - static, R - RIP, M - mobile, B - BGP
       D - EIGRP, EX - EIGRP external, O - OSPF, IA - OSPF inter area
       N1 - OSPF NSSA external type 1, N2 - OSPF NSSA external type 2
       E1 - OSPF external type 1, E2 - OSPF external type 2
       i - IS-IS, su - IS-IS summary, L1 - IS-IS level-1, L2 - IS-IS level-2
       ia - IS-IS inter area, * - candidate default, U - per-user static route
       o - ODR, P - periodic downloaded static route

Gateway of last resort is not set

         100.0.0.0/24 is subnetted, 1 subnets
C   100.1.1.0 is directly connected, Serial2/0
         172.17.0.0/24 is subnetted, 2 subnets
R   172.17.1.0 [120/1] via 172.17.2.1, 00:00:24, FastEthernet1/0
C   172.17.2.0 is directly connected, FastEthernet1/0
         172.16.0.0/24 is subnetted, 2 subnets
O   172.16.1.0 [110/2] via 172.16.2.1, 00:04:21, FastEthernet0/0
C   172.16.2.0 is directly connected, FastEthernet0/0

R1#show ip route

```
Codes: C - connected, S - static, R - RIP, M - mobile, B - BGP
       D - EIGRP, EX - EIGRP external, O - OSPF, IA - OSPF inter area
       N1 - OSPF NSSA external type 1, N2 - OSPF NSSA external type 2
       E1 - OSPF external type 1, E2 - OSPF external type 2
       i - IS-IS, su - IS-IS summary, L1 - IS-IS level-1, L2 - IS-IS level-2
ia - IS-IS inter area, * - candidate default, U - per-user static route
       o - ODR, P - periodic downloaded static route
Gateway of last resort is not set
       172.16.0.0/24 is subnetted, 2 subnets
C      172.16.1.0 is directly connected, FastEthernet0/0
C      172.16.2.0 is directly connected, FastEthernet1/0
R2#show ip route
Codes: C - connected, S - static, R - RIP, M - mobile, B - BGP
       D - EIGRP, EX - EIGRP external, O - OSPF, IA - OSPF inter area
       N1 - OSPF NSSA external type 1, N2 - OSPF NSSA external type 2
       E1 - OSPF external type 1, E2 - OSPF external type 2
       i - IS-IS, su - IS-IS summary, L1 - IS-IS level-1, L2 - IS-IS level-2
ia - IS-IS inter area, * - candidate default, U - per-user static route
       o - ODR, P - periodic downloaded static route
Gateway of last resort is not set
       172.17.0.0/24 is subnetted, 2 subnets
C      172.17.1.0 is directly connected, FastEthernet0/0
C      172.17.2.0 is directly connected, FastEthernet1/0
```

步骤五：实施路由重分布（OSPF 与 RIP）。
（1）在 R0 上设置路由重分布。

```
R0(config)#router rip
R0(config-router)#redistribute ospf 1 metric 4          //将 OSPF 路由条目到 RIP，度量值为 4
R0(config-router)#exit
R0(config)#router ospf 1
R0(config-router)#redistribute rip subnets metric 10    //将 RIP 路由条目到 OSPF，度量值为 10
R0(config-router)#
```

注意：RIP 以跳数作为度量值参数，RIP 有效跳数为 15 跳，OSPF 默认度量值为 20。
（2）查看 R1 和 R2 的路由表。

```
R1#show ip route
Codes: C - connected, S - static, R - RIP, M - mobile, B - BGP
       D - EIGRP, EX - EIGRP external, O - OSPF, IA - OSPF inter area
       N1 - OSPF NSSA external type 1, N2 - OSPF NSSA external type 2
```

E1 - OSPF external type 1, E2 - OSPF external type 2
i - IS-IS, su - IS-IS summary, L1 - IS-IS level-1, L2 - IS-IS level-2
ia - IS-IS inter area, * - candidate default, U - per-user static route
o - ODR, P - periodic downloaded static route

Gateway of last resort is not set

```
     172.17.0.0/24 is subnetted, 2 subnets
O E2    172.17.1.0 [110/10] via 172.16.2.2, 00:03:04, FastEthernet1/0
O E2    172.17.2.0 [110/10] via 172.16.2.2, 00:03:04, FastEthernet1/0
     172.16.0.0/24 is subnetted, 2 subnets
C       172.16.1.0 is directly connected, FastEthernet0/0
C       172.16.2.0 is directly connected, FastEthernet1/0
```

注意：E2 表示来自外部区域的路由条目。

```
R2#show ip route
Codes: C - connected, S - static, R - RIP, M - mobile, B - BGP
       D - EIGRP, EX - EIGRP external, O - OSPF, IA - OSPF inter area
       N1 - OSPF NSSA external type 1, N2 - OSPF NSSA external type 2
       E1 - OSPF external type 1, E2 - OSPF external type 2
       i - IS-IS, su - IS-IS summary, L1 - IS-IS level-1, L2 - IS-IS level-2
       ia - IS-IS inter area, * - candidate default, U - per-user static route
       o - ODR, P - periodic downloaded static route

Gateway of last resort is not set

     172.17.0.0/24 is subnetted, 2 subnets
C       172.17.1.0 is directly connected, FastEthernet0/0
C       172.17.2.0 is directly connected, FastEthernet1/0
     172.16.0.0/24 is subnetted, 2 subnets
R       172.16.1.0 [120/4] via 172.17.2.2, 00:00:08, FastEthernet1/0
R       172.16.2.0 [120/4] via 172.17.2.2, 00:00:08, FastEthernet1/0
```

步骤六：在 R0 上实施默认路由并重分布到 RIP 和 OSPF 区域。

（1）重分布静态到 RIP 和 OSPF 区域。

```
R0(config)#ip route 0.0.0.0 0.0.0.0 100.1.1.2
R0(config)#router rip
R0(config-router)#redistribute static metric 4            //重分布静态路由
R0(config-router)#exit
R0(config-router)#default-information originate always    //重分布静态路由
```

注意：
① 在 RIP 中重分布静态路由的度量值参数可以省略；
② always 参数可以省略。

（2）查看 R1 和 R2 路由表。

R1#show ip route
Codes: C - connected, S - static, R - RIP, M - mobile, B - BGP
       D - EIGRP, EX - EIGRP external, O - OSPF, IA - OSPF inter area
       N1 - OSPF NSSA external type 1, N2 - OSPF NSSA external type 2
       E1 - OSPF external type 1, E2 - OSPF external type 2
       i - IS-IS, su - IS-IS summary, L1 - IS-IS level-1, L2 - IS-IS level-2
ia - IS-IS inter area, * - candidate default, U - per-user static route
       o - ODR, P - periodic downloaded static route
Gateway of last resort is 172.16.2.2 to network 0.0.0.0
       172.17.0.0/24 is subnetted, 2 subnets
O E2   172.17.1.0 [110/10] via 172.16.2.2, 00:16:25, FastEthernet1/0
O E2   172.17.2.0 [110/10] via 172.16.2.2, 00:16:25, FastEthernet1/0
       172.16.0.0/24 is subnetted, 2 subnets
C   172.16.1.0 is directly connected, FastEthernet0/0
C   172.16.2.0 is directly connected, FastEthernet1/0
O*E2 0.0.0.0/0 [110/1] via 172.16.2.2, 00:04:40, FastEthernet1/0
R2#show ip route
Codes: C - connected, S - static, R - RIP, M - mobile, B - BGP
       D - EIGRP, EX - EIGRP external, O - OSPF, IA - OSPF inter area
       N1 - OSPF NSSA external type 1, N2 - OSPF NSSA external type 2
       E1 - OSPF external type 1, E2 - OSPF external type 2
       i - IS-IS, su - IS-IS summary, L1 - IS-IS level-1, L2 - IS-IS level-2
ia - IS-IS inter area, * - candidate default, U - per-user static route
       o - ODR, P - periodic downloaded static route
Gateway of last resort is 172.17.2.2 to network 0.0.0.0
       172.17.0.0/24 is subnetted, 2 subnets
C   172.17.1.0 is directly connected, FastEthernet0/0
C   172.17.2.0 is directly connected, FastEthernet1/0
       172.16.0.0/24 is subnetted, 2 subnets
R   172.16.1.0 [120/4] via 172.17.2.2, 00:00:05, FastEthernet1/0
R   172.16.2.0 [120/4] via 172.17.2.2, 00:00:05, FastEthernet1/0
R*   0.0.0.0/0 [120/4] via 172.17.2.2, 00:00:05, FastEthernet1/0

步骤七：在 ISP 路由实施默认路由实现内外网互通。

ISP(config)#ip route 0.0.0.0 0.0.0.0 100.1.1.1

在 PC1、PC2 和 Server 上正确配置 IP 地址信息，测试内外网连通性，PC1 均能与 PC2 和 Server 通信。

7. 总　结

在实施路由重分布过程中应注意路由环路，路由器有可能把从一个自治系统学到的路由信息发送回自治系统，特别是在做双向重分布的时候，每一种路由协议的度量标准不同，所有路由器通过重分布所选择的路径可能并非最佳路径，应考虑路由信息的兼容性问题；每种路由协议的工作原理不一样，故路由协议收敛的时间也不一致。

项目 17 虚拟路由器冗余协议（VRRP）配置

1. 工程目标

（1）理解 VRRP 技术原理。

（2）掌握 VRRP 配置。

2. 技术要点

虚拟路由器冗余协议（Virtual Router Redundancy Protocol，VRRP）的工作原理和 HSRP 非常类似，不过 VRRP 是国际标准，允许在不同厂商的设备之间运行。VRRP 由一组路由器组成单一的虚拟路由器。VRRP 中虚拟网关的地址可以和接口上的地址相同，VRRP 中的接口只有三个状态：初始状态（Initial）、主状态（Master）和备份状态（Backup）。

VRRP 根据优先级来确定备份组中每台路由器的角色（Master 路由器或 Backup 路由器）。优先级越高，则越有可能成为 Master 路由器。VRRP 优先级的取值范围为 0~255（数值越大表明优先级越高），可配置的范围是 1~254。优先级 0 有特殊的意义，它表示当前的主路由器已经不参与 VRRP 的运作，将 VRRP 中某个路由的物理地址设置为 VRRP 组的地址，这个路由器的优先级为 255。

3. 设备清单

路由器（3 台）；网络连线（若干根）；测试 PC（1 台）。

4. 工作场景

如图 2.11 所示模拟的是某网络中为了实现路由的冗余，由 R2 和 R3 分别担任主路由和次路由设备，当 R2 或者 R2 上的链路出现问题时，PC 可以通过 R3 访问 R1 的环回口。

5. 地址规划

两个分散校园园区网络地址规划如表 2.9 所示。

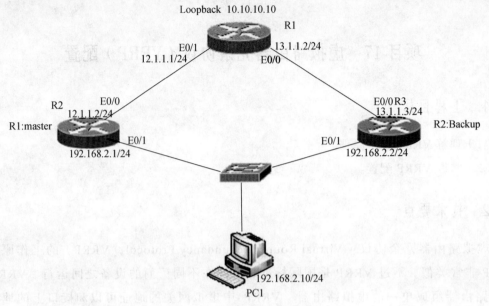

图 2.11 区域网络工作场景

表 2.9 两个分散校园园区网络地址规划

设备名称	设备及端口的配置地址		备注
R1	E0/0	13.1.1.1/24	
	E0/1	12.1.1.1/24	
	Loopback 0	1.1.1.1/24	
R2	E0/0	12.1.1.2/24	
	E0/1	192.168.2.1/24	
R3	E0/0	13.1.1.3/24	
	E0/1	192.168.2.2/24	
PC1	192.168.2.10/24		

6. 工作过程

步骤一：连接设备。按图 2.11 使用网络线缆在工作现场连接好设备。

步骤二：配置路由器设备基本接口信息及路由协议，保证路由间能够相互连通。

（1）配置 R1 路由器端口的地址。

R1#configure terminal

R1(config)#int loopback 0

R1(config-if)#ip add 1.1.1.1 255.255.255.0

R1(config-if)#exit

R1(config)#int e0/1

R1(config-if)#ip add 12.1.1.1 255.255.255.0
R1(config-if)#no shutdown
R1(config-if)#exit
R1(config)#int e0/0
R1(config-if)#ip add 13.1.1.1 255.255.255.0
R1(config-if)#no shutdown
R1(config-if)#exit
R1(config)#router rip
R1(config-router)#version 2
R1(config-router)#no auto-summary
R1(config-router)#network 1.1.1.0
R1(config-router)#network 12.1.1.0
R1(config-router)#network 13.1.1.0
R1(config-router)#end

（2）配置 R2 路由器的端口地址。

R2#configure termianal
R2(config)#int e0/0
R2(config-if)#ip add 12.1.1.2 255.255.255.0
R2(config-if)#no shutdown
R2(config-if)#exit
R2(config)#int e0/1
R2(config-if)#ip add 192.168.2.1 255.255.255.0
R2(config-if)#no shutdown
R2(config-if)#exit
R2(config)#router rip
R2(config-router)#version 2
R2(config-router)#no auto-summary
R2(config-router)#network 12.0.0.0
R2(config-router)#network 192.168.2.0
R2(config-router)#end
R2#wr

（3）配置 R3 路由器的端口地址。

R3#configure terminal
R3(config)#int e0/0
R3(config-if)#ip add 13.1.1.3 255.255.255.0
R3(config-if)#no shutdown
R3(config-if)#exit

R3(config)#int e0/1
R3(config-if)#ip add 192.168.2.2 255.255.255.0
R3(config-if)#no shutdown
R3(config-if)#exit
R3(config)#router rip
R3(config-router)#version 2
R3(config-router)#no auto-summary
R3(config-router)#network 13.0.0.0
R3(config-router)#network 192.168.2.0
R3(config-router)#end
R3#wr

（4）配置 VRRP 协议。

R2(config)#int e0/1
R2(config-if)#vrrp 1 ip 192.168.2.1
R2(config-if)#vrrp 1 priority 200
R2(config-if)#vrrp 1 preempt
R2(config-if)#end
R2#show vrrp
Ethernet0/1 - Group 1
 State is Master
 Virtual IP address is 192.168.2.1
 Virtual MAC address is 0000.5e00.0101
 Advertisement interval is 1.000 sec
 Preemption enabled
 Priority is 255 (cfgd 200)
 Master Router is 192.168.2.1 (local), priority is 255
 Master Advertisement interval is 1.000 sec
 Master Down interval is 3.003 sec

R3(config)#int e0/1
R3(config-if)#vrrp 1 ip 192.168.2.1
R3(config-if)#vrrp 1 priority 110
R3(config-if)#vrrp 1 preempt
R3(config-if)#end
R3#wr
R3#show vrrp
Ethernet0/1 - Group 1
 State is Backup

Virtual IP address is 192.168.2.1

　　Virtual MAC address is 0000.5e00.0101

　　Advertisement interval is 1.000 sec

　　Preemption enabled

　　Priority is 110

　　Master Router is 192.168.2.1, priority is 255

　　Master Advertisement interval is 1.000 sec

（5）测试 VRRP 协议，关闭 R2 的 E0/1 接口，查看 R2 和 R3 状态。

R2(config)#int e0/1

R2(config-if)#shut

R2(config-if)#shutdown

R2(config-if)#end

R2#show vrrp

Ethernet0/1 - Group 1

　　State is Init

　　Virtual IP address is 192.168.2.1

　　Virtual MAC address is 0000.5e00.0101

　　Advertisement interval is 1.000 sec

　　Preemption enabled

　　Priority is 255 (cfgd 200)

　　Master Router is unknown, priority is unknown

　　Master Advertisement interval is unknown

　　Master Down interval is unknown

R3#show vrrp

Ethernet0/1 - Group 1

　　State is Master

　　Virtual IP address is 192.168.2.1

　　Virtual MAC address is 0000.5e00.0101

　　Advertisement interval is 1.000 sec

　　Preemption enabled

　　Priority is 110

　　Master Router is 192.168.2.2 (local), priority is 110

　　Master Advertisement interval is 1.000 sec

　　Master Down interval is 3.570 sec

7. 总　结

VRRP 能够实现路由备份，达到链路稳定可靠的目的，配置过程首先要保证网络是连通的，然后开启 VRRP 配置，通过调整路由器的优先级确定哪个路由器承担主路由。可以通过在主路由和备份路由上开启 debug 命令，在图 2.11 中用 PC1 ping R1 的环回口，然后断开 R2 的链路，再次从 PC1 ping R1 的环回口，观察数据包通过 R2 或 R3 的情况。

第 3 章　网络安全实践篇

项目 18　配置标准访问列表控制网络流量

1. 工程目标

（1）理解标准 ACL 的作用。
（2）理解标准 ACL 的定义。
（3）掌握标准 ACL 的应用。

2. 技术要点

ACL 为访问控制列表，它使用包过滤技术，在路由器上读取第三层及第四层包头中的信息，如源地址、目的地址、源端口、目的端口等，根据预先定义好的规则对包进行过滤，从而达到访问控制的目的。

（1）标准 ACL。

标准 ACL 最简单，是基于 IP 包中的源地址进行过滤，编号范围为 1~99 或 1 300~1 999。

（2）扩展 ACL。

扩展 ACL 比标准 ACL 具有更多的匹配项，功能更加强大和细化，可以根据包协议类型、源地址、目的地址、源端口、目的端口、TCP 连接建立等进行过滤，编号范围为 100~199 或 2 000~2 699。

（3）名称 ACL。

以列表名称代替列表编号来定义 ACL，包括标准 ACL 和扩展 ACL。

（4）通配符掩码。

通配符掩码（wildcard-mask）路由器使用的通配符掩码（或反掩码）与源或目标地址一起来分辨匹配的地址范围。它跟子网掩码刚好相反，不同于子网掩码告诉路由器 IP 地址的哪一位属于网络号，通配符掩码告诉路由器为了判断出匹配，它需要检查 IP 地址中的多少位。这个地址掩码可以只使用两个 32 位的号码来确定 IP 地址的范围。两种特殊的通配符掩码是"255.255.255.255"和"0.0.0.0"，前者等价于关键字"any"，而后者等价于关键字"host"。

（5）Inbound 和 outbound。

当在接口上应用访问控制列表时，用户要明确访问控制列表是应用于流入数据还是流出数据。

3. 设备清单

路由器（2台）；网络连线（若干根）；测试PC（2台）。

4. 工作场景

如图3.1所示，网络拓扑是信息大学计算机科学技术学院学生网和行政办公网网络工作场景，要实现学生网（192.168.3.0）和教师办公网（192.168.1.0）的隔离，可以在其中R1路由器上做标准ACL技术控制，以实现网络之间的隔离。

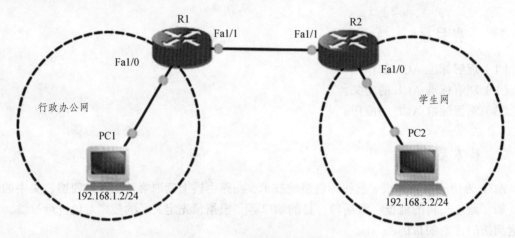

图3.1　区域网络工作场景

5. 地址规划

信息大学计算机科学技术学院学生网和行政办公网网络地址的规划过程如表3.1所示。

表3.1　校园网地址规划

设备名称	设备及端口的配置地址		备 注
R1	Fa1/0	192.168.1.1/24	局域网端口，连接PC1
	Fa1/1	192.168.2.1/24	局域网端口，连接R2路由器Fa1/1
R2	Fa1/1	192.168.2.2/24	局域网端口，连接R1路由器Fa1/1
	Fa1/0	192.168.3.1/24	局域网端口，连接PC2
PC1	192.168.1.2/24		网关：192.168.1.1
PC2	192.168.3.2/24		网关：192.168.3.1

注：在实际的实验室方案中，也可以使用两台R1完成任务。如果实验室没有第二台R2设备，可以使用三层交换机代替路由器。

6. 工作过程

步骤一：连接设备。使用准备好的网线，按照图 3.1 所示网络拓扑连接好设备。

步骤二：配置行政办公网络的路由器 R1。

使用配置 PC 通过 Console 端口连接路由器 R1，进入路由器 R1 的配置模式状态，配置路由器 R1 信息：端口 IP 地址和动态 RIP 路由，从而实现网络连通。

R1#
R1#configure terminal
R1(config)#interface fa1/0
R1(config-if)#ip address 192.168.1.1 255.255.255.0
R1(config-if)#no shutdown
R1(config-if)#interface fa1/1
R1(config-if)#ip address 192.168.2.1 255.255.255.0
R1(config-if)#no shutdown
R1(config-if)#exit
R1(config)#router rip
R1(config-router)#network 192.168.1.0
R1(config-router)#network 192.168.2.0
R1(config-router)#exit
R1(config)#

步骤三：配置学生网络的路由器 R2。

使用配置 PC 通过 Console 端口连接路由器 R2，进入路由器 R2 的配置模式，配置路由器 R2 信息：端口 IP 地址和动态 RIP 路由，从而实现网络连通。

R2(config)#interface fa1/1
R2(config-if)#ip address 192.168.2.2 255.255.255.0
R2(config-if)#no shutdown
R2(config-if)#interface fa1/0
R2(config-if)#ip address 192.168.3.1 255.255.255.0
R2(config-if)#no shutdown
R2(config-if)#exit
R2(config)#router rip
R2(config-router)#network 192.168.2.0
R2(config-router)#network 192.168.3.0
R2(config-router)#exit
R2(config)#

步骤四：测试从学生网到行政办公网的连通性。

（1）使用测试计算机 PC1 和 PC2，分别代表行政办公网（192.168.1.0）和学生网

（192.168.3.0）中的两台设备，分别为它们配置相应网段的地址信息。

（2）使用 ping 命令测试从学生网到行政办公网的连通性，网络连通正常。

步骤五：禁止学生网访问行政办公网。

因为禁止来源于一个网络的数据流，按照 ACL 配置控制规则，应该选择标准的 ACL 技术解决方案。按照标准的 ACL 应用规则，尽量把数据流限制在离目标网络近的地点，以尽可能扩大源网络访问的范围，因此选择接近目标网络的路由器 R1 启用安全策略。

（1）在路由器 R1 上配置标准 ACL 控制规则。

R1(config)#access-list 10 deny 192.168.3.0 0.0.0.255
R1(config)#access-list 10 permit any
R1(config)#
R1(config)#interface fa1/0
R1(config-if)# ip access-group 10 out
R1(config-if)#exit
R1(config)#

（2）在路由器 R1 的 Fa1/0 端口上使用编制好的 ACL 控制规则。

步骤六：给 PC1 和 PC2 配置好 IP 地址信息，测试 PC1 与 PC2 连通性，测试结果是 PC1 ping 不通 PC2。

7. 总 结

根据减少不必要通信流量的通行准则，标准 ACL 设置比较简单，只根据源地址制定策略，一般规则应用在离目的主机近的路由器上做。

项目 19 配置扩展访问列表保护服务器安全

1. 工程目标

（1）理解扩展 ACL 的作用。
（2）掌握标准 ACL 的应用。

2. 技术要点

扩展 IP 访问列表可以根据数据包的源 IP、目的 IP、源端口、目的端口、协议来定义规则，进行数据包的过滤。

在路由器上应用 ACL 的一般规则为：可以为每种协议（per protocol）、每个方向（per direction）、每个接口（per interface）配置一个 ACL。

每种协议一个 ACL：要控制接口上的流量，必须为接口上启用的每种协议定义相应的 ACL。

每个方向一个 ACL：一个 ACL 只能控制接口上一个方向的流量。要控制入站流量和出站流量，必须分别定义两个 ACL。

每个接口一个 ACL：一个 ACL 只能控制一个接口（例如 Fast Ethernet 0/0）上的流量。

ACL 的编写可能相当复杂而且极具挑战性。在每个接口上都可以针对多种协议和各个方向进行定义。

3. 设备清单

路由器（2 台）；网络连线（若干根）；测试 PC（2 台）。

4. 工作场景

如图 3.2 所示网络拓扑是信息大学计算机科学技术学院学生网和教师网网络工作场景，要实现教师网（192.168.1.0）和学生网（192.168.3.0）之间的互相连通，但不允许学生网访问教师网中的 FTP 服务器，可以在路由器 R2 上做扩展 ACL 技术控制，以实现网络之间的隔离。

5. 地址规划

信息大学计算机科学技术学院学生网和教师网网络地址的规划过程如表 3.2 所示。

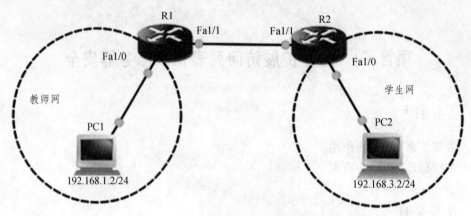

图 3.2 区域网络工作场景

表 3.2 校园网地址规划

设备名称	设备及端口的配置地址		备 注
R1	Fa1/0	192.168.1.1/24	局域网端口，连接 PC1
	Fa1/1	192.168.2.1/24	局域网端口，连接 R2 路由器 Fa1/1
R2	Fa1/1	192.168.2.2/24	局域网端口，连接 R1 路由器 Fa1/1
	Fa1/0	192.168.3.1/24	局域网端口，连接 PC2
PC1	192.168.1.2/24		网关：192.168.1.1
PC2	192.168.3.2/24		网关：192.168.3.1

注：在实际的实验室方案中，也可以使用两台 R1 完成任务。如果实验室没有第二台 R2 设备，可以使用三层交换机代替路由器。

6. 工作过程

步骤一：连接设备。使用准备好的网线，按照图 3.2 所示网络拓扑连接好设备。
步骤二：配置教师网络的路由器 R1。

使用配置 PC 通过 Console 端口连接路由器 R1，进入路由器 R1 的配置模式状态，配置路由器 R1 信息：端口 IP 地址和动态 RIP 路由，从而实现网络连通。

R1#
R1#configure terminal
R1(config)#interface fa1/0
R1(config-if)#ip address 192.168.1.1 255.255.255.0
R1(config-if)#no shutdown
R1(config-if)#interface fa1/1
R1(config-if)#ip address 192.168.2.1 255.255.255.0
R1(config-if)#no shutdown
R1(config-if)#exit
R1(config)#router rip

R1(config-router)#network 192.168.1.0

R1(config-router)#network 192.168.2.0

R1(config-router)#exit

R1(config)#

步骤三：配置学生网络的路由器 R2。

使用配置 PC 通过 Console 端口连接路由器 R2，进入路由器 R2 的配置模式，配置路由器 R2 信息：端口 IP 地址和动态 RIP 路由，从而实现网络连通。

R2(config)#interface fa1/1

R2(config-if)#ip address 192.168.2.2 255.255.255.0

R2(config-if)#no shutdown

R2(config-if)#interface fa1/0

R2(config-if)#ip address 192.168.3.1 255.255.255.0

R2(config-if)#no shutdown

R2(config-if)#exit

R2(config)#router rip

R2(config-router)#network 192.168.2.0

R2(config-router)#network 192.168.3.0

R2(config-router)#exit

R2(config)#

步骤四：测试从学生网到行政办公网的连通性。

（1）使用测试计算机 PC1 和 PC2，分别代表教师网（192.168.1.0）和学生网（192.168.3.0）中的两台设备，分别为它们配置相应网段的地址信息。

（2）使用 ping 命令测试从学生网到教师网的连通性，网络连通正常。

步骤五：禁止学生网访问教师网中的 FTP 服务器。

因为是可以访问目标网络，但禁止访问目标网络的某一项服务的数据流，按照 ACL 配置分类规则，应该选择使用扩展 ACL 控制技术。

按照扩展 ACL 应用规则，应尽量把数据流限制在离发源网络近的地点，尽可能减少从源网络流出的无效数据流占用的网络带宽，因此选择接近源头网络的路由器 R2 启用安全策略。

（1）在路由器 R2 上配置扩展 ACL 控制规则。

R2(config)# access-list 101 deny tcp 172.16.3.0 0.0.0.255 172.16.1.0 0.0.0.255 eq ftp

R2 (config)#access-list 101 permit ip any any

R2 (config)#

（2）在路由器 R2 的 Fa1/0 端口上使用编制好的 ACL 控制规则。

R2 (config)#interface fa1/0

R2 (config-if)#ip access-group 101 in

R2 (config-if)#no shutdown

R2 (config-if)#exit

R2 (config)#

（3）使用ping命令测试网络连通性。由于使用了扩展ACL技术，ACL控制通过路由器R2上的数据流，禁止学生网访问教师网中的FTP服务器，PC2能ping通PC1，但不能访问PC1上的FTP服务器。

7. 总　结

ACL配置一般规则：

（1）每个接口、每个协议或每个方向只可以应用一个访问列表。ACL末尾都隐含拒绝的语句，经过第一个ACL的过滤，不符合的包都被丢弃，也就不会留下任何包和第二个ACL比较。

（2）除非在ACL末尾有Permit any命令，否则所有和列表条件不符合的包都将丢弃，所以每个ACL至少要有一个运行语句，以免其拒绝所有流量。

（3）要先创建ACL，再将其应用到一个接口上，才会生效。

（4）ACL过滤通过路由器的流量，但不过滤该路由器产生的流量。

（5）IP扩展访问列表尽可能的应用在靠近源地址的接口上，因为它可以基于目的地址、协议等过滤，放在源端过滤，避免需要过滤的数据包被路由到目的端才被过滤，以节省带宽。

第 4 章　IP 地址服务实践篇

项目 20　静态 NAT 配置

1. 工程目标

（1）掌握 NAT 的工作原理。
（2）掌握 NAT 的应用特点。
（3）掌握静态 NAT 的工作过程。
（4）掌握静态 NAT 的配置方法。

2. 技术要点

1）概　述

网络地址转换（Network Address Translation，NAT）的功能就是指在一个网络内部根据需要随意自定义 IP 的地址，而不需要经过申请。在网络内部，各计算机间通过内部的 IP 地址进行通信。而当内部的计算机要与外部 Internet 网络进行通信时，具有 NAT 功能的设备（如路由器）负责将其内部的 IP 地址转换为合法的 IP 地址（即经过申请的 IP 地址）进行通信。

2）NAT 的应用环境

情况 1：一个企业不想让外部网络用户知道自己的网络内部结构，可以通过 NAT 将内部网络与外部 Internet 隔离开，因此外部用户根本不知道通过 NAT 设置的内部 IP 地址。

情况 2：一个企业申请的合法 Internet IP 地址很少，而内部网络用户很多。可以通过 NAT 功能实现多个用户同时共用一个合法 IP 与外部 Internet 进行通信。

3）NAT 分类

① 静态地址转换。

静态地址转换将内部本地地址与内部合法地址进行一对一地转换，且需要指定和哪个合法地址进行转换。如果内部网络有 WWW 服务器或 FTP 服务器等可以为外部用户提供服务，则这些服务器的 IP 地址必须采用静态地址转换，以便外部用户可以使用这些服务。

② 动态地址转换。

动态地址转换也是将内部本地地址与内部合法地址一对一地转换，但是动态地址转换是从内部合法地址池中动态地选择一个未使用的地址来对内部本地地址进行转换的。

③ 复用动态地址转换。

复用动态地址转换首先是一种动态地址转换，但是它可以允许多个内部本地地址共用一个内部合法地址。对只申请到少量 IP 地址但却经常同时有多个用户上外部网络的情况，这种转换极为有用。过载配置（Ooverload）就是符合条件数据的源地址（N 个，例如：局域网的各个私有 IP）都被转换为一个合法 IP，或者多个合法 IP。

4）术 语

内部本地地址（Inside localaddress）：分配给内部网络中的计算机的内部 IP 地址。

内部合法地址（Inside globaladdress）：对外进行 IP 通信时，代表一个或多个内部本地地址的合法 IP 地址。需要申请才可取得的 IP 地址。

5）静态 NAT 工作原理

静态转换是指将内部网络的私有 IP 地址转换为公有 IP 地址，IP 地址对是一对一的，是一成不变的，即某个私有 IP 地址只转换为某个公有 IP 地址。借助于静态转换，可以实现外部网络对内部网络中某些特定设备（如服务器）的访问。

静态 NAT 使用本地地址与全局地址的一对一映射，且这些映射保持不变。静态 NAT 对于必须具有一致的地址、可从 Internet 访问的 Web 服务器或主机特别有用。这些内部主机可能是企业服务器或网络设备。

6）静态地址转换基本配置过程

① 在内部本地地址与内部合法地址之间建立静态地址转换。在全局设置状态下输入：

Ip nat inside source static 内部本地地址 内部合法地址

② 指定连接网络的内部端口，在端口设置状态下输入：

ip nat inside

③ 指定连接外部网络的外部端口，在端口设置状态下输入：

ip nat outside

注：可以根据实际需要定义多个内部端口及多个外部端口。

3. 设备清单

路由器（2 台）；交换机（1 台）；网络连线（若干根）；测试 PC（3 台）。

4. 工作场景

某单位创建了 PC1 和 PC2，这两台 PC 机不但允许内部用户（IP 地址为 172.16.1.0/24 网段）进行访问，而且要求 Internet 上的外网用户也能够进行访问。为实现此功能，本单位向当地的 ISP 申请了一段公网的 IP 地址 200.1.1.0/24，通过静态 NAT 转换，当 Internet 上的用户访问这两台 PC 时，实际访问的是 200.1.1.11 和 200.1.1.12 这两个公网的 IP 地址，但用户的访问数据被路由器 Router-A 分别转换为 172.16.1.11 和 172.16.1.12 两个内网的私有 IP 地址，网络拓扑图如图 4.1 所示。

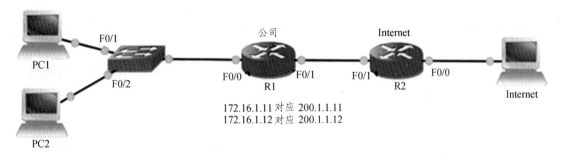

图 4.1 区域网络的工作场景

5. 地址规划

某单位网络地址的规划信息如表 4.1 所示。

表 4.1 校园网地址规划

设备名称	设备及端口的配置地址		备 注
R1	F0/0	172.16.1.1/24	局域网端口，连接 PC1
	F0/1	200.1.1.2/24	广域网端口，连接 R2 路由器 F0/1
R2	F0/1	200.1.1.1/24	广域网端口，连接 R1 路由器 F0/1
	F0/0	192.168.1.1/24	局域网端口，连接 Internet
PC1	172.16.1.11/24		网关：172.16.1.1
PC2	172.16.1.12/24		网关：172.16.1.1
Internet	192.168.1.10		网关：192.168.1.1

6. 工作过程

步骤一：连接设备。使用准备好的网线按照图 4.1 所示网络拓扑连接好设备。
步骤二：配置公司网络的路由器 R1。
使用配置 PC 通过 Console 端口连接路由器 R1，进入路由器 R1 的配置模式状态，配置路由器 R1 信息：端口 IP 地址，动态 RIP 路由，实现网络连通。

```
R1#configure terminal                          //进入全局配置模式
R1(config)#in f0/1                             //进入端口 F0/1
R1(config-if)#ip address 200.1.1.2 255.255.255.0   //配置 IP 地址
R1(config-if)#no shut                          //启用端口，使其转发数据
R1(config-if)#exit
```

R1(config)#in f0/0   //进入端口 F0/0
R1(config-if)#ip address 172.16.1.1 255.255.255.0   //配置 IP 地址
R1(config-if)#no shutdown   //打开端口
R1(config-if)#exit
R1(config)#ip route 0.0.0.0 0.0.0.0 f0/1   //设置以端口为 F0/1 的默认静态路由到外网

步骤三：路由器 R2 的基本配置。

R2#configure terminal
R2 (config)#in f0/0
R2 (config-if)#ip address 192.168.1.1 255.255.255.0
R2 (config-if)#exit
R2 (config)#in f0/1
R2 (config-if)#ip address 200.1.1.1 255.255.255.0
R2 (config-if)#no shut
R2 (config-if)#exit
R2 (config)#ip route 0.0.0.0 0.0.0.0 f0/1

步骤四：在路由器 R1 上配置静态 NAT。

R1(config）# interface fastethernet 0/0
R1(config-if）#ip nat inside
R1(config-if）#exit
R1(config）# interface fastethernet 0/1
R1(config-if）#ip nat outside
R1(config-if）#exit
R1(config）#ip nat inside source static 172.16.1.11 200.1.1.11
//将内网的 172.16.1.11IP 地址静态映射为外网的 200.1.1.11 公有 IP 地址
R1（config）#ip nat inside source static 172.16.1.12 200.1.1.12
//将内网的 172.16.1.12IP 地址静态映射为外网的 200.1.1.12 公有 IP 地址
R1（config）#end

注意：不要把 inside 和 outside 弄错。

步骤五：测试。

在路由器 R2 的 FastEthernet0/0 端口上接入一台 PC，IP 地址设置为 192.168.1.2，网关为 172.16.1.1（路由器 R2 上 FastEthernet0/0 端口的 IP 地址），然后进入"命令提示符"窗口，用 ping 命令分别测试 200.1.1.11 和 200.1.1.12 两个 IP 地址，应该是连通的。说明路由器 R1 已进行了地址转换。其实，真正访问的是 172.16.1.11 和 172.16.1.12 两台主机。如图 4.2 所示。

在路由器 Router-A 上使用 show ip nat translations 命令查看 NAT 的转换情况。验证信息：

PC1>ping 192.168.1.10

```
Pinging 192.168.1.10 with 32 bytes of data:

Reply from 192.168.1.10: bytes=32 time=125ms TTL=126
Reply from 192.168.1.10: bytes=32 time=110ms TTL=126
Reply from 192.168.1.10: bytes=32 time=125ms TTL=126
Reply from 192.168.1.10: bytes=32 time=110ms TTL=126

Ping statistics for 192.168.1.10:
    Packets: Sent = 4, Received = 4, Lost = 0 (0% loss),
Approximate round trip times in milli-seconds:
    Minimum = 110ms, Maximum = 125ms, Average = 117ms
```

图 4.2　测试网络连通性

Router#show ip nat translations　　//显示活动的转换
Pro　　Inside global　　Inside local　　Outside local　　Outside global
Icmp　　200.1.1.11:17　172.16.1.11:17　192.168.1.10:17　192.168.1.10:17
Icmp　　200.1.1.11:18　172.16.1.11:18　192.168.1.10:18　192.168.1.10:18
icmp　　200.1.1.11:19　172.16.1.11:19　192.168.1.10:19　192.168.1.10:19

7. 总　　结

静态 NAT 为内部地址与外部地址的一对一映射。静态 NAT 允许外部设备发起与内部设备的连接。配置静态 NAT 转换很简单，首先需要定义要转换的地址，然后在适当的接口上配置 NAT。从指定的 IP 地址到达内部接口的数据包需经过转换。外部接口收到的以指定 IP 地址为目的地的数据包也需经过转换。

项目 21 端口多路复用（PAT）配置

1. 工程目标

（1）掌握 PAT 的工作原理。
（2）掌握 PAT 的应用特点。
（3）掌握静态 PAT 的工作过程。
（4）掌握静态 PAT 的配置方法。

2. 技术要点

1）PAT 工作原理

端口多路复用（Port Address Translation，PAT）是指改变外出数据包的源端口并进行端口转换，即端口地址转换（Port Address Translation，PAT）。采用端口多路复用方式，内部网络的所有主机均可共享一个合法外部 IP 地址实现对 Internet 的访问，从而可以最大限度地节约 IP 地址资源。同时，又可隐藏网络内部的所有主机，有效避免来自 Internet 的攻击。因此，目前网络中应用最多的就是端口多路复用方式。

2）PAT 地址转换过程

（1）在全局设置模式下，定义内部合地址池。
ip nat pool　地址池名字起始　IP 地址终止　IP 地址子网掩码
（2）在全局设置模式下，定义一个标准的 access-list 规则以允许哪些内部本地地址可以进行动态地址转换。
access-list　标号　permit 源地址通配符
其中，标号为 1~99 之间的整数。
（3）在全局设置模式下，设置在内部的本地地址与内部合法 IP 地址间建立复用动态地址转换。
ip nat inside source list　访问列表标号　pool　内部合法地址池名字　overload
（4）在端口设置状态下，指定与内部网络相连的内部端口。
ip nat inside
（5）在端口设置状态下，指定与外部网络相连的外部端口。
ip nat outside

3. 设备清单

路由器（2 台）；交换机（1 台）；网络连线（若干根）；测试 PC（3 台）。

4. 工作场景

现假设某单位创建了 PC1 和 PC2,这两台 PC 机不但允许内部用户(IP 地址为 172.16.1.0/24 网段)进行访问,而且要求 Internet 上的外网用户也能够进行访问。为实现此功能,本单位向当地的 ISP 申请了一段公网的 IP 地址 200.1.1.0/24,通过 PAT 转换,当 Internet 上的用户访问这两台 PC 时,实际访问的是 200.1.1.11 和 200.1.1.12 这两个公网的 IP 地址,但用户的访问数据被路由器 Router-A 分别转换为 172.16.1.11 和 172.16.1.12 两个内网的私有 IP 地址。网络拓扑图如图 4.3 所示。

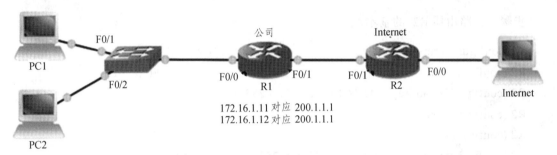

图 4.3　区域网络的工作场景

5. 地址规划

某单位网络地址的规划信息如表 4.2 所示。

表 4.2　校园网地址规划

设备名称	设备及端口的配置地址		备　注
R1	F0/0	172.16.1.1/24	局域网端口,连接 PC1
	F0/1	200.1.1.1/24	广域网端口,连接 R2 路由器 F0/1
R2	F0/1	200.1.1.2/24	广域网端口,连接 R1 路由器 F0/1
	F0/0	192.168.1.1/24	局域网端口,连接 Internet
PC1		172.16.1.11/24	网关:172.16.1.1
PC2		172.16.1.12/24	网关:172.16.1.1
Internet		192.168.1.10	网关:192.168.1.1

6. 工作过程

步骤一:连接设备。使用准备好的网线,按照图 4.3 所示网络拓扑连接好设备。
步骤二:配置公司网络的路由器 R1。

使用配置 PC 通过 Console 端口连接路由器 R1,进入路由器 R1 的配置模式状态,配置路由器 R1 信息:端口 IP 地址,动态 RIP 路由,实现网络连通。

```
R1#configure terminal                              //进入全局配置模式
R1(config)#in f0/1                                 //进入端口 F0/1
R1(config-if)#ip address 200.1.1.1 255.255.255.0   //配置 IP 地址
```

R1(config-if)#no shut	//启用端口，使其转发数据
R1(config-if)#exit	
R1(config)#in f0/0	//进入端口 F0/0
R1(config-if)#ip address 172.16.1.1 255.255.255.0	//配置 IP 地址
R1(config-if)#no shutdown	//打开端口
R1(config-if)#exit	
R1(config)#ip route 0.0.0.0 0.0.0.0 f0/1	//设置以端口为 F0/1 的默认静态路由到外网

步骤三：路由器 R2 的基本配置。

R2#configure terminal	//进入全局配置模式
R2 (config)#in f0/0	//进入端口 F0/0
R2 (config-if)#ip address 192.168.1.1 255.255.255.0	//配置 IP 地址
R2 (config-if)#exit	
R2 (config)#in f0/1	//进入端口 F0/1
R2 (config-if)#ip address 200.1.1.2 255.255.255.0	//配置 IP 地址
R2 (config-if)#no shut	//启用端口，使其转发数据
R2 (config-if)#exit	
R2 (config)#ip route 0.0.0.0 0.0.0.0 f0/1	//设置以端口为 F0/1 的默认静态路由

步骤四：在路由器 R1 上配置静态 NAT。

R1(config)# interface fastethernet 0/0	//进入端口 F0/0
R1(config-if)#ip nat inside	//将 F0/0 端口定义为内部端口
R1(config-if)#exit	
R1(config)# interface fastethernet 0/1	//进入端口 F0/1
R1(config-if)#ip nat outside	//将 F0/1 端口定义为外部端口
R1(config-if)#exit	
R1(config)#access-list 1 permit 172.16.1.0 0.0.0.255	//定义内部网允许通过的 IP 地址
R1(config)#ip nat pool wl 200.1.1.1 200.1.1.1 netmask 255.255.255.0	//定义公有地址池范围
R1(config)#ip nat inside source list 1 pool wl overload	//建立公有地址与私有地址的映射关系

注意：

① 如果主机的数量不是很多，可以直接使用 outside 接口地址配置 PAT，不必定义公有地址池。

② R1（config）#ip nat inside source list 1 interface f0/1 overload //定义内部源地址控制列表的送出端口

③ 不要把 inside 和 outside 弄错。

步骤五：测试。

PC1 的 IP 地址设置为 172.16.1.11，网关为 172.16.1.1（路由器 R1 上 FastEthernet0/0 端口

第 4 章 IP 地址服务实践篇

的 IP 地址），再 PC1 上进入"命令提示符"窗口，用 ping 命令分别测试 192.168.1.1 和 192.168.1.10 两个 IP 地址，应该是连通的，说明路由器 R1 已进行了地址转换，如图 4.4 所示。

```
Pinging 192.168.1.10 with 32 bytes of data:

Reply from 192.168.1.10: bytes=32 time=125ms TTL=126
Reply from 192.168.1.10: bytes=32 time=110ms TTL=126
Reply from 192.168.1.10: bytes=32 time=125ms TTL=126
Reply from 192.168.1.10: bytes=32 time=110ms TTL=126

Ping statistics for 192.168.1.10:
    Packets: Sent = 4, Received = 4, Lost = 0 (0% loss),
Approximate round trip times in milli-seconds:
    Minimum = 110ms, Maximum = 125ms, Average = 117ms
```

图 4.4 连通性测试

在路由器 Router-A 上使用 show ip nat translations 命令查看 NAT 的转换情况。验证信息：
PC1>ping 192.168.1.10
Router#show ip nat translations   //显示活动的转换
Pro   Inside global   Inside local   Outside local   Outside global
Icmp 200.1.1.1:17 172.16.1.11:17 192.168.1.10:17   192.168.1.10:17
Icmp 200.1.1.1:18 172.16.1.11:18   192.168.1.10:18   192.168.1.10:18
icmp 200.1.1.1:19   172.16.1.11:19   192.168.1.10:19   192.168.1.10:19

7. 总　　结

PAT 是把内部地址映射到外部网络的 IP 地址的不同端口上，从而可以实现多对一的映射。不仅完美地解决了 IP 地址不足的问题，而且还能够有效避免来自网络外部的攻击，隐藏并保护网络内部的计算机。

项目 22　动态主机配置协议（DHCP）配置

1. 工程目标

（1）掌握 DHCP 的工作原理。
（2）掌握在交换机、路由器上 DHCP 的配置方法。

2. 技术要点

动态主机配置协议（Dynamic Host Configuration Protocol，DHCP）是一个局域网的网络协议，使用 UDP 协议工作，管理员可以利用 DHCP 服务器，从预先设置的 IP 地址池中，自动给主机分配 IP 地址，这样不仅能保证 IP 地址不重复分配，也能及时回收 IP 地址，从而提高 IP 地址的利用率。

DHCP 客户机在启动时，会搜寻网络中是否存在 DHCP 服务器。如果找到，则给 DHCP 服务器发送一个请求。DHCP 服务器接到请求后，为 DHCP 客户机选择 TCP/IP 配置的参数，并把这些参数发送给客户端。如果已配置冲突检测设置，则 DHCP 服务器在将租约中的地址提供给客户机之前会试用 ping 测试作用域中每个可用地址的连通性。这可确保提供给客户的每个 IP 地址都没有被使用手动 TCP/IP 配置的另一台非 DHCP 计算机使用。

① 寻找 DHCP 服务器。

当使用 DHCP 客户端第一次登录网络的时候，计算机发现本机上没有任何 IP 地址设定，将以广播方式发送 DHCP discover 发现信息来寻找 DHCP 服务器，网络上每一台安装了 TCP/IP 协议的主机都会接收这个广播信息，但只有 DHCP 服务器才会做出响应。

② 分配 IP 地址。

在网络中接收到 DHCP discover 发现信息的 DHCP 服务器都会做出相应，它从尚未分配的 IP 地址中挑选一个分配给 DHCP 客户机，向 DHCP 客户机发送一个包含分配的 IP 地址和其他设置的 DHCP offer 提供信息。

③ 接受 IP 地址。

DHCP 客户端接受到 DHCP offer 提供信息之后，选择第一个接收到的提供信息，然后以广播的方式回答一个 DHCP request 请求信息，该信息包含向它所选定的 DHCP 服务器请求 IP 地址的内容。

④ IP 地址分配确认。

当 DHCP 服务器收到 DHCP 客户端回答的 DHCP request 请求信息之后，便向 DHCP 客户端发送一个包含它所提供的 IP 地址和其他设置的 DHCP ack 确认信息，告诉 DHCP 客户端可以使用它提供的 IP 地址。然后，DHCP 客户机便将其 TCP/IP 协议与网卡绑定，另外，除了 DHCP 客户机选中的服务器外，其他的 DHCP 服务器将收回曾经提供的 IP 地址。

3. 设备清单

三层交换机（1台）；二层交换机（1台）；网络连线（若干根）；测试PC（2台）。

4. 工作场景

某个只有40人规模的公司，该公司建设有局域网，有2个逻辑网络为VLAN2（市场部）和VLAN3（行政部），为了节约网络建设成本，单位决定将DHCP服务器架设在汇聚层交换机，给公司各终端设备自动分配IP地址。具体网络拓扑如图4.5所示。

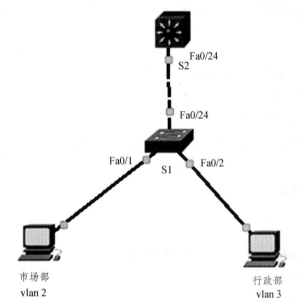

图 4.5　区域网络的工作场景

5. 地址规划

某单位网络地址的规划信息如表4.3所示。

表 4.3　校园网地址规划

设备名称	设备及端口的配置地址		备 注
S2	vlan 2	172.16.1.1/24	虚拟局域网端口，连接市场部
	vlan 3	172.16.2.1/24	虚拟局域网端口，连接行政部
PC1	自动获取地址		网关：172.16.1.1（自动获取）
PC2	自动获取地址		网关：172.16.2.1（自动获取）

6. 工作过程

步骤一：连接设备。使用准备好的网线，按照图4.5所示网络拓扑连接好设备。

步骤二：配置公司网络的三层交换机 S2 和二层交换机 S1。

使用配置 PC 通过 Console 端口连接交换机 S1、S2，进入路由器 S1 的配置模式状态，配置 S1 的 VLAN 信息。

（1）交换机的配置过程。

```
S1(config)#vlan 2                              //新建 VLAN
S1(config)#vlan 3
S1(config)#interface fastEthernet 0/1
S1(config-if)#switchport mode access
S1(config-if)#switchport access vlan 2         //将接口划分到 VLAN
S1(config)#interface fastEthernet 0/2
S1(config-if)#switchport mode access
S1(config-if)#switchport access vlan 3
S1(config)#interface fastEthernet 0/24
S1(config-if)#switchport mode trunk
```

（2）三层交换的配置过程，利用 SVI 实现 VLAN 间通信。

```
S2(config)#vlan 2
S2(config-vlan)#vlan 3
S2(config-vlan)#exit
S2(config)#interface vlan 2
S2(config-if)#ip address 172.16.1.1 255.255.255.0
S2(config-if)#exit
S2(config)#interface vlan 3
S2(config-if)#ip address 172.16.2.1 255.255.255.0
S2(config-if)#exit
S2(config)#ip routing
```

（3）在 S2 上配置 DHCP 服务器。

```
S2(config)#ip dhcp excluded-address 172.16.1.1 172.16.1.10   //指定排除地址
S2(config)#ip dhcp pool shichangbu                            //给地址池命名
S2(dhcp-config)#network 172.16.1.0 255.255.255.0              //设置地址池范围
S2(dhcp-config)#dns-server 119.6.6.6                          //地址 DNS 地址
S2(dhcp-config)#default-router 172.16.1.1                     //指定网关地址
S2(dhcp-config)#exit
S2(config)#ip dhcp excluded-address 172.16.2.1
S2(config)#ip dhcp pool xingzhengbu
S2(dhcp-config)#network 172.16.2.0 255.255.255.0
S2(dhcp-config)#default-router 172.16.2.1
S2(dhcp-config)#dns-server 119.6.6.6
```

（4）测试。

在 PC1、PC2 上设置 IP 地址获取的方式选择自动获得 IP 地址，如图 4.6~4.7 所示。此方式适合于同网段 DHCP 自动分配地址，适用范围较小，DHCP 服务器功能利用率不高。

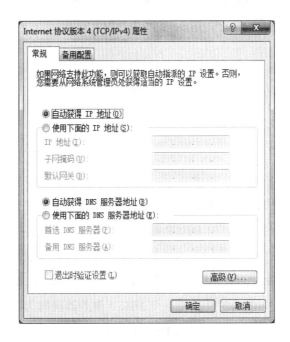

图 4.6　客户端地址设置

```
PC>ipconfig /all
Physical Address............: 00D0.9769.2E52
IP Address..................: 172.16.1.11
Subnet Mask.................: 255.255.255.0
Default Gateway.............: 172.16.1.1
DNS Servers.................: 119.6.6.6
```

图 4.7　客户端地址获取信息

7．总　　结

在路由器或三层交换机搭建 DHCP 服务器一般适合于小型局域网，不适合分布在不同路由器和交换机上 VLAN 主机获取地址，大型局域网应在专用的服务器上部署 DHCP 服务器。

项目 23　DHCP 中继配置

1. 工程目标

（1）掌握 DHCP 的工作原理。

（2）掌握 DHCP 中继配置方法。

2. 技术要点

企业组网时，需要根据实际来划分不同的 VLAN。在 DHCP 工作过程中发现，DHCP 客户使用 IP 广播来寻找同一网段上的 DHCP 服务器。当服务器和客户段处在不同网段，即被路由器分割开来时，路由器是不会转发这样的广播包的。因此可能需要在每个网段上设置一个 DHCP 服务器，虽然 DHCP 只消耗很小的一部分资源，但多个 DHCP 服务器，毕竟要带来管理上的不方便。DHCP 中继的使用使一个 DHCP 服务器同时为多个网段服务成为可能，路由器可以帮助转发广播请求数据包到达 DHCP 服务器。

3. 设备清单

三层交换机（1台）；二层交换机（1台）；网络连线（若干根）；测试 PC（2台）。

4. 工作场景

某个只有 100 人规模的公司，该公司建设有局域网，有两个逻辑网络为 VLAN2（市场部）和 VLAN3（行政部），公司有独立的 DHCP 服务器，为给公司各终端设备自动分配 IP 地址。具体网络拓扑如图 4.8 所示。

5. 地址规划

某单位网络地址的规划信息如表 4.4 所示。

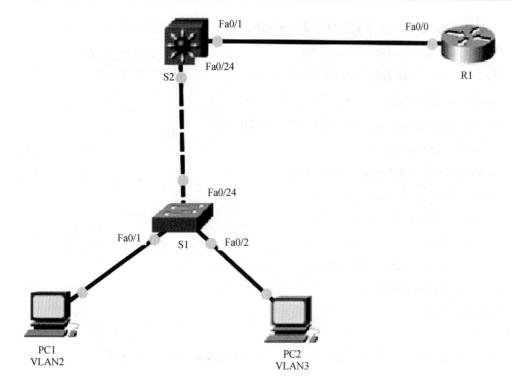

图 4.8 区域网络的工作场景

表 4.4 校园网地址规划

设备名称	设备及端口的配置地址		备 注
S2	vlan 2	172.16.1.1/24	虚拟局域网端口，连接市场部
	vlan 3	172.16.2.1/24	虚拟局域网端口，连接行政部
	Fa0/1	192.168.1.1/24	与 R1 的 Fa0/0 相连
R1	Fa0/0	192.168.1.2/24	与 S2 的 Fa0/1 相连
PC1	自动获取地址		网关：172.16.1.1（自动获取）
PC2	自动获取地址		网关：172.16.2.1（自动获取）

6. 工作过程

步骤一：连接设备。使用准备好的网线，按照图 4.8 所示网络拓扑连接好设备。
步骤二：配置公司网络的三层交换机 S2 路由器 R1。
（1）R1 路由器的配置过程。

R1(config)#ip dhcp pool student   //给地址池命名

```
R1(dhcp-config)#network 172.16.1.0 255.255.255.0      //设置地址池范围
R1(dhcp-config)#dns-server 119.6.6.6                  //地址 DNS 地址
R1(dhcp-config)#default-router 172.16.1.1             //指定网关地址
R1(dhcp-config)#exit
R1(config)#ip dhcp excluded-address 172.16.2.1
R1(config)#ip dhcp pool teacher
R1(dhcp-config)#network 172.16.2.0 255.255.255.0
R1(dhcp-config)#default-router 172.16.2.1
R1(dhcp-config)#dns-server 119.6.6.6
R1(config)#interface fastEthernet 0/0
R1(config-if)#ip address 192.168.1.2 255.255.255.0
R1(config-if)#no shutdown
R1(config-if)#exit
R1(config)#router ospf 1                              //启用 OSPF 协议让全网通信
R1(config-router)#network 192.168.1.0 0.0.0.255 area 0   //宣告直连网段
```

（2）三层交换 S2 的配置过程。

```
S2(config)#interface vlan 2
S2(config-if)#ip address 172.16.1.1 255.255.255.0     //给 VLAN 指定 IP 地址
S2(config)#interface vlan 3
S2(config-if)#ip address 172.16.2.1 255.255.255.0
S2(config)#interface fastEthernet 0/1
S2(config-if)#no switchport                           //改变接口类型（由二层接口
                                                        修改为三层接口）
S2(config-if)#ip address 192.168.1.1 255.255.255.0    //配置 IP 地址
S2(config)#ip routing                                 //开启三层交换机路由功能
S2(config)#router ospf 1                              //启用 OSPF 协议让
S2(config-router)#network 192.168.1.0 0.0.0.255 area 0
S2(config-router)#network 172.16.1.0 0.0.0.255 area 0
S2(config-router)#network 172.16.2.0 0.0.0.255 area 0
S2(config)#interface vlan 2
S2(config-if)#ip helper-address 192.168.1.2           //配置 DHCP 中继转发给 DHCP
                                                        服务器
S2(config)#interface vlan 3
S2(config-if)#ip helper-address 192.168.1.2
```

二层交换机 S1 配置与项目 22 同。

（3）测试。

设置 PC1、PC2 地址获取方式为自动获得 IP 地址，两台 PC 均能获得 IP 地址，此时方式适用范围较广，DHCP 服务器在网内较少，方便管理，适用性较强。

7. 总　　结

DHCP 是一个优秀的 IP 地址管理工具，减少了由于手工设置可能出现的错误，并极大地提高了工作效率，降低了劳动强度，同时也便于网络管理员的维护。当网络使用的 IP 地址段改变了，管理员只需修改 DHCP 服务器的 IP 地址池即可，而不必逐台修改网络内的所有计算机地址。

第 5 章 综合实践篇

项目 24 中小企业园区网建设

1. 工程目标

通过项目实施,掌握企业网络建设的需求,并给出解决方案实施。

2. 技术要点

(1) VLAN 配置与管理;
(2) 单臂路由技术;
(3) 静态路由;
(4) 动态路由协议;
(5) 路由重分布;
(6) NAT 技术。

3. 设备清单

路由器(3台);三层交换机(2台),交换机(2台);V35线缆(1条);网络连线(若干根);测试 PC(8台),WWW 服务器(1台)。

4. 工作场景

某企业计划建设自己的企业园区网络,希望通过新建的网络,提供一个安全、可靠、可扩展、高效的网络环境。该企业有两个办公地点,且距离较远,其中总公司部门较多,有业务部、财务部、行政部等,为主要办公场所,对交换网络可靠性和可用性的要求较高。分公司办公人员较少,只有综合部和项目部。公司内部上 Internet 通过 NAT 技术实现。具体网络拓扑如图 5.1 所示。

5. 地址规划

某单位网络地址的规划过程如表 5.1 所示。

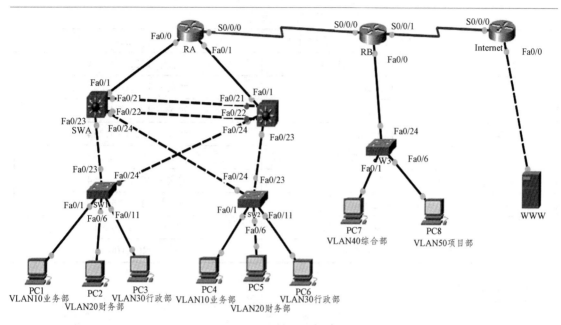

图 5.1　区域网络的工作场景

表 5.1　校园网地址规划

设备名称	设备及端口的配置地址		备注
Internet	F0/0	200.100.2.1/24	外网服务器，连接 WWW 服务器
	S0/0/1	200.100.1.2/24	广域网端口，连接 RB S0/0/1
RA	Fa0/0	172.16.10.2/24	局域网端口，连接 SWA
	Fa0/1	172.16.20.2/24	局域网端口，连接 SWB
	S0/0/0	172.16.30.1/24	广域网端口，连接 RB 路由器 S0/0/0
RB	S0/0/0	172.16.30.2/24	广域网端口，连接 RA 路由器 S0/0/0
	S0/0/1	200.100.1.1/24	广域网端口，连接 Internet S0/0/1
SWA	F0/1	172.16.10.1/24	
	VLAN 10	172.16.1.1/24	
	VLAN 20	172.16.2.1/24	
	VLAN 30	172.16.3.1/24	
SWB	F0/1	172.16.20.1/24	
	VLAN 10	172.16.1.2/24	
	VLAN 20	172.16.2.2/24	
	VLAN 30	172.16.3.2/24	
	F0/1.40	172.16.4.1/24	
	F0/1.50	172.16.5.1/24	

设备名称	设备及端口的配置地址	备 注
PC1	172.16.1.11/24	网关：172.16.1.254
PC2	172.16.2.11/24	网关：172.16.2.254
PC3	172.16.3.11/24	网关：172.16.1.254
PC4	172.16.1.12/24	网关：172.16.1.254
PC5	172.16.2.12/24	网关：172.16.1.254
PC6	172.16.3.12/24	网关：172.16.1.254
PC7	172.16.4.2/24	网关：172.16.1.1
PC8	172.16.5.2/24	网关：172.16.5.1
WWW	200.100.2.2	网关：200.100.2.1

6. 工作过程

步骤一：连接设备。使用准备好的网线，按照图5.1所示网络拓扑连接好设备。

步骤二：配置公司网络的5台交换机上建VLAN10、VLAN20、VLAN30、VLAN40、VLAN40。

SW1(config)#vlan 10
SW1(config-vlan)#name yewubu
SW1(config-vlan)#vlan 20
SW1(config-vlan)#name caiwubu
SW1(config-vlan)#vlan 30
SW1(config-vlan)#name xingzhengbu
SW2(config)#vlan 10
SW2(config-vlan)#name yewubu
SW2(config-vlan)#vlan 20
SW2(config-vlan)#name caiwubu
SW2(config-vlan)#vlan 30
SW2(config-vlan)#name xingzhengbu

SWA(config)#vlan 10
SWA(config-vlan)#name yewubu
SWA(config-vlan)#vlan 20
SWA(config-vlan)#name caiwubu
SWA(config-vlan)#vlan 30
SWA(config-vlan)#name xingzhengbu

SWB(config)#vlan 10
SWB(config-vlan)#name yewubu
SWB(config-vlan)#vlan 20
SWB(config-vlan)#name caiwubu
SWB(config-vlan)#vlan 30
SWB(config-vlan)#name xingzhengbu

SW3(config)#vlan 40
SW3(config-vlan)#name zonghebu
SW3(config-vlan)#vlan 50
SW3(config-vlan)#name xiangmubu

步骤三：在 SW1 和 SW2 上将 1～5 端口，6～10 端口、11～15 端口分别划到 VLAN 10、VLAN 20、VLAN 30 中，在 SW3 将 1～5 端口，6～10 端口分别划到 VLAN 40、VLAN 50 中。

SW1(config)#interface range fastEthernet 0/1-5
SW1(config-if-range)#switchport mode access
SW1(config-if-range)#switchport access vlan 10
SW1(config-if-range)#exit
SW1(config)#interface range fastEthernet 0/6-10
SW1(config-if-range)#switchport mode access
SW1(config-if-range)#switchport access vlan 20
SW1(config-if-range)#exit
SW1(config)#interface range fastEthernet 0/11-15
SW1(config-if-range)#switchport mode access
SW1(config-if-range)#switchport access vlan 30

SW2(config)#interface range fastEthernet 0/1-5
SW2(config-if-range)#switchport mode access
SW2(config-if-range)#switchport access vlan 10
SW2(config-if-range)#exit
SW2(config)#interface range fastEthernet 0/6-10
SW2(config-if-range)#switchport mode access
SW2(config-if-range)#switchport access vlan 20
SW2(config-if-range)#exit
SW2(config)#interface range fastEthernet 0/11-15
SW2(config-if-range)#switchport mode access
SW2(config-if-range)#switchport access vlan 30

SW3(config)#interface range fastEthernet 0/1-5

SW3(config-if-range)#switchport mode access
SW3(config-if-range)#switchport access vlan 40
SW3(config-if-range)#exit
SW3(config)#interface range fastEthernet 0/6-10
SW3(config-if-range)#switchport mode access
SW3(config-if-range)#switchport access vlan 50

步骤四：把交换机 SW1、SW2 和 SW3 与上联 SWA、SWB 和 RB 的端口设置为中继模式。

SWA(config)#interface range fastEthernet 0/23-24
SWA(config-if-range)#switchport mode trunk

SWB(config)#interface range fastEthernet 0/23-24
SWB(config-if-range)#switchport mode trunk

SW1(config)#interface range fastEthernet 0/23-24
SW1(config-if-range)#switchport mode trunk

SW2(config)#interface range fastEthernet 0/23-24
SW2(config-if-range)#switchport mode trunk

Sw3(config)#interface fastEthernet 0/24
Sw3(config-if)#switchport mode trunk

步骤五：把两台三层交换机之间的 f0/21 和 f0/22 端口配置聚合端口。

SWA(config)#interface range fastEthernet 0/21-22
SWA(config-if-range)#channel-group 1 mode on
SWA(config)#interface port-channel 1
SWA(config-if)#switchport mode trunk

SWB(config)#interface range fastEthernet 0/21-22
SWB(config-if-range)#channel-group 1 mode on
SWB(config)#interface port-channel 1
SWB(config-if)#switchport mode trunk

步骤六：在交换机 SWA、SWB、SW1 和 SW2 上配置生成树，指定 SWA 为根桥，SWB 为备份根桥。

SWA(config)#spanning-tree vlan 10 root primary
SWA(config)#spanning-tree vlan 10 priority 4096
SWA(config)#spanning-tree vlan 20 root primary

SWA(config)#spanning-tree vlan 20 priority 4096
SWA(config)#spanning-tree vlan 30 root secondary
SWA(config)#spanning-tree vlan 30 priority 8192

SWB(config)#spanning-tree vlan 10 root secondary
SWB(config)#spanning-tree vlan 10 priority 8192
SWB(config)#spanning-tree vlan 20 root secondary
SWB(config)#spanning-tree vlan 20 priority 8192
SWB(config)#spanning-tree vlan 30 root primary
SWB(config)#spanning-tree vlan 30 priority4096

步骤七：在三层交换机上配置 SVI 实现 VLAN 间通信，同时在 RB 上做单臂路由实现 VLAN 40 与 VLAN 50 通信。

SWA(config)#interface vlan 10
SWA(config-if)#ip address 172.16.1.1 255.255.255.0
SWA(config-if)#exit
SWA(config)#interface vlan 30
SWA(config-if)#ip address 172.16.2.1 255.255.255.0
SWA(config-if)#exit
SWA(config)#interface vlan 30
SWA(config-if)#ip address 172.16.3.1 255.255.255.0
SWA(config-if)#exit
SWA(config)#ip routing
SWB(config)#interface vlan 10
SWB(config-if)#ip address 172.16.1.2 255.255.255.0
SWB(config-if)#exit
SWB(config)#interface vlan 30
SWB(config-if)#ip address 172.16.2.2 255.255.255.0
SWB(config-if)#exit
SWB(config)#interface vlan 30
SWB(config-if)#ip address 172.16.3.2 255.255.255.0
SWB(config-if)#exit
SWB(config)#ip routing

RB(config)#interface fastEthernet 0/0.40
RB(config-subif)#encapsulation dot1Q 40
RB(config-subif)#ip address 172.16.4.1 255.255.255.0
RB(config-subif)#exit
RB(config)#interface fastEthernet 0/0.50

RB(config-subif)#encapsulation dot1Q 50
RB(config-subif)#ip address 172.16.5.1 255.255.255.0

步骤八：在 SWA 和 SWB 上做 VRRP 实现冗余网关功能。

SWA(config)#interface vlan10
SWA(config-if)#vrrp 1 priority 120
SWA(config-if)#vrrp 1 ip 172.16.1.254
SWA(config)#interface vlan20
SWA(config-if)#vrrp 2 priority 120
SWA(config-if)#vrrp 2 ip 172.16.2.254
SWA(config)#interface vlan30
SWA(config-if)#vrrp 3 ip 172.16.1.254

SWB(config)#interface vlan10
SWB(config-if)#vrrp 1 ip 172.16.1.254
SWB(config)#interface vlan20
SWB(config-if)#vrrp 2 ip 172.16.2.254
SWBconfig)#interface vlan30
SWB(config-if)#vrrp 3 priority 120
SWB(config-if)#vrrp 3 ip 172.16.1.254

步骤九：在三层交换的路由端口、RA 和 RB 和 Internet 的路由上配置接口 IP 地址。

SWA(config)#interface fastEthernet 0/1
SWA(config-if)#no switchport
SWA(config-if)#ip address 172.16.10.1 255.255.255.0

SWB (config)#interface fastEthernet 0/1
SWB(config-if)#no switchport
SWB(config-if)#ip address 172.16.20.1 255.255.255.
RA(config)#interface fastEthernet 0/0
RA(config-if)#no shutdown
RA(config-if)#ip address 172.16.10.2 255.255.255.0
RA(config-if)#exit
RA(config)#interface fastEthernet 0/1
RA(config-if)#no shutdown
RA(config-if)#ip address 172.16.20.2 255.255.255.0
RA(config-if)#exit
RA(config)#interface serial 0/0/0
RA(config-if)#no shutdown

RA(config-if)#ip address 172.16.30.1 255.255.255.0

RB(config)#interface serial 0/0/0
RB(config-if)#no shutdown
RB(config-if)#ip address 172.16.30.2 255.255.255.0
RB(config-if)#exit
RB(config)#interface serial 0/0/1
RB(config-if)#no shutdown
RB(config-if)#ip address 200.100.1.1 255.255.255.0

Internet(config)#interface serial 0/0/1
Internet(config-if)#no shutdown
Internet(config-if)#ip address 200.100.1.2 255.255.255.0
Internet(config-if)#exit
Internet(config)#interface fastEthernet 0/0
Internet(config-if)#no shutdown
Internet(config-if)#ip address 200.100.2.1 255.255.255.0

步骤十：在企业网内部启用 OSPF 路由协议实现内网通信，在 RB 路由器上启用默认路由实现企业内网到互联网的访问。

SWA(config)#router ospf 1
SWA(config-router)#router-id 1.1.1.1
SWA(config-router)#network 172.16.1.0 0.0.0.255 a
SWA(config-router)#network 172.16.1.0 0.0.0.255 area 0
SWA(config-router)#network 172.16.2.0 0.0.0.255 area 0
SWA(config-router)#network 172.16.3.0 0.0.0.255 area 0
SWA(config-router)#network 172.16.10.0 0.0.0.255 area 0
SWB(config)#router ospf 1
SWB(config-router)#router-id 2.2.2.2
SWB(config-router)#network 172.16.1.0 0.0.0.255 a
SWB(config-router)#network 172.16.1.0 0.0.0.255 area 0
SWB(config-router)#network 172.16.2.0 0.0.0.255 area 0
SWB(config-router)#network 172.16.3.0 0.0.0.255 area 0
SWB(config-router)#network 172.16.20.0 0.0.0.255 area 0

RA(config)#router ospf 1
RA(config-router)#router-id 3.3.3.3
RA(config-router)#network 172.16.10.0 0.0.0.255 a
RA(config-router)#network 172.16.10.0 0.0.0.255 area 0

RA(config-router)#network 172.16.20.0 0.0.0.255 area 0
RA(config-router)#network 172.16.30.0 0.0.0.255 area 0

RB(config)#router ospf 1
RB(config-router)#router-id 4.4.4.4
RB(config-router)#network 172.16.30.0 0.0.0.255 area 0
RB(config-router)#exit
RB(config)#ip route 0.0.0.0 0.0.0.0 200.100.1.2

步骤十一：路由重分布，将 RB 上的默认路由重分布到内网的 OSPF 协议中。

RB(config)#router ospf 1
RB(config-router)#default-information o
RB(config-router)#default-information originate

步骤十二：在 RB 上做 NAT 实现内网对外网的访问，可用的公有地址为 200.100.1.1。

RB(config)#access-list 10 permit 172.16.0.0 0.0.255.255
RB(config)#ip nat pool qiye 200.100.1.1 200.100.1.1 netmask 255.255.255.0
RB(config)#ip nat inside source list 10 pool qiye overload
RB(config)#interface serial 0/0/0
RB(config-if)#ip nat inside
RB(config-if)#exit
RB(config)#interface fastEthernet 0/0.40
RB(config-subif)#ip nat inside
RB(config-subif)#exit
RB(config)#interface fastEthernet 0/0.50
RB(config-subif)#ip nat inside
RB(config-subif)#exit
RB(config)#interface serial 0/0/1
RB(config-if)#ip nat outside

步骤十二：在内网主机和 WWW 服务器上配置正确的 IP 地址信息进行测试。

文轨车书　交通天下
http://www.xnjdcbs.com

责任编辑　李芳芳
特邀编辑　韩迎春　林莉
封面设计　米想Miga Graphic design studio

- Android项目开发实训
- 综合布线实训
- 移动网络分析与优化
- 计算机维护与服务规范——7天精通PC维护
- 基于.NET平台的Web开发
- Visual C#程序设计基础
- 计算机网络技术
- 企业级Java EE商业项目开发
- 企业案例软件测试技术
- 网络互联实践
- 项目化静态网页设计简明教程

ISBN 978-7-5643-4242-5

定价：20.00元

数学建模简明教程

柏宏斌 兰恒友 陈德勤 ○ 主编

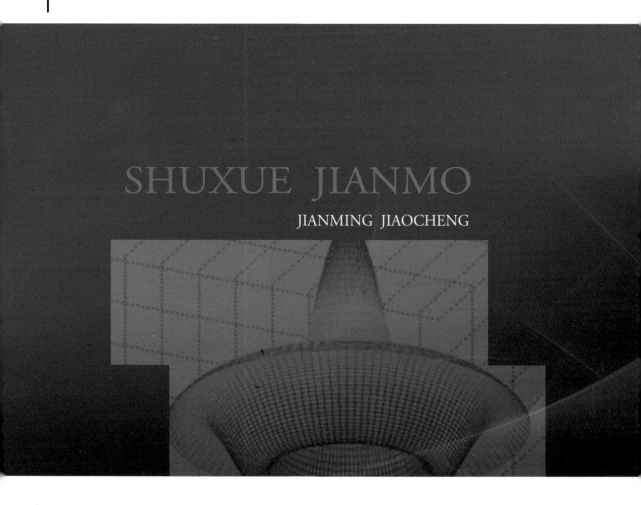